Synthesis Lectures on Engineering, Science, and Technology

The focus of this series is general topics, and applications about, and for, engineers and scientists on a wide array of applications, methods and advances. Most titles cover subjects such as professional development, education, and study skills, as well as basic introductory undergraduate material and other topics appropriate for a broader and less technical audience.

Ghada Alsuhli · Vasilis Sakellariou ·
Hani Saleh · Mahmoud Al-Qutayri ·
Baker Mohammad · Thanos Stouraitis

Number Systems for Deep Neural Network Architectures

 Springer

Ghada Alsuhli
Khalifa University
Abu Dhabi, United Arab Emirates

Vasilis Sakellariou
Khalifa University
Abu Dhabi, United Arab Emirates

Hani Saleh
Khalifa University
Abu Dhabi, United Arab Emirates

Mahmoud Al-Qutayri
Khalifa University
Abu Dhabi, United Arab Emirates

Baker Mohammad
Khalifa University
Abu Dhabi, United Arab Emirates

Thanos Stouraitis
Khalifa University
Abu Dhabi, United Arab Emirates

ISSN 2690-0300   ISSN 2690-0327 (electronic)
Synthesis Lectures on Engineering, Science, and Technology
ISBN 978-3-031-38132-4   ISBN 978-3-031-38133-1 (eBook)
https://doi.org/10.1007/978-3-031-38133-1

© The Editor(s) (if applicable) and The Author(s), under exclusive license to Springer Nature
Switzerland AG 2024

This work is subject to copyright. All rights are solely and exclusively licensed by the Publisher, whether the whole
or part of the material is concerned, specifically the rights of translation, reprinting, reuse of illustrations, recitation,
broadcasting, reproduction on microfilms or in any other physical way, and transmission or information storage
and retrieval, electronic adaptation, computer software, or by similar or dissimilar methodology now known or
hereafter developed.
The use of general descriptive names, registered names, trademarks, service marks, etc. in this publication does
not imply, even in the absence of a specific statement, that such names are exempt from the relevant protective
laws and regulations and therefore free for general use.
The publisher, the authors, and the editors are safe to assume that the advice and information in this book are
believed to be true and accurate at the date of publication. Neither the publisher nor the authors or the editors give
a warranty, expressed or implied, with respect to the material contained herein or for any errors or omissions that
may have been made. The publisher remains neutral with regard to jurisdictional claims in published maps and
institutional affiliations.

This Springer imprint is published by the registered company Springer Nature Switzerland AG
The registered company address is: Gewerbestrasse 11, 6330 Cham, Switzerland

Acknowledgements

This work was supported by the Khalifa University of Science and Technology under Award CIRA-2020-053.

Contents

Acronyms

ADA	Add-decode-accumulate
AI	Artificial intelligence
ASICs	Application-specific integrated circuit
BFP	Block floating point
CLBs	Configurable logic blocks
CNNs	Convolutional neural networks
CPUs	Central processing units
DFXP	Dynamic fixed point
DNNs	Deep neural networks
ELU	Extended linear unit
FLP	Floating point
FPGAs	Field programmable gate arrays
FPU	Floating point unit
FXP	Fixed point
GPUs	Graphics processing units
K-L	Kullback–Leibler
LNS	Logarithmic number system
LSTM	Long short-term memory
LUTs	Look-up tables
MAC	Multiply accumulate
NLP	Natural Language Processing
PNS	Posit Number System
ReLU	Rectified linear unit
RNNs	Recurrent neural networks
SOC	System-on-chip
Unum	Universal number

Introduction

1

Abstract

In this introductory chapter, we provide an overview of the main topics covered in this book and the motivations to write it. The importance of efficient number systems for Deep Neural Networks (DNNs) and their impact on hardware design and performance are emphasized. In addition, we list the various number systems that will be discussed in detail in the subsequent chapters. Finally, we outline the organization of the book with a summary of the contents of each chapter, to offer readers a clear roadmap of what to expect while exploring number systems for DNNs in this book.

1.1 Introduction to Number Systems for DNNs

During the past decade, DNNs have shown outstanding performance in a myriad of Artificial Intelligence (AI) applications. Since their success in both speech [1] and image recognition [2], great attention has been drawn to DNNs from academia and industry, which subsequently led to a wide range of products that utilize them [3]. Although DNNs are inspired by the deep hierarchical structures of the human brain, they have exceeded human accuracy in a number of domains [4]. Nowadays, the contribution of DNNs is notable in many fields including self-driving cars [5], speech recognition [6], computer vision [7], natural language processing [8], and medical applications [9]. This DNN revolution is helped by the massive accumulation of data and the rapid growth in computing power [10].

Due to the substantial computational complexity and memory demands, accelerating DNN processing has typically relied on either high-performance general-purpose compute engines like Central Processing Units (CPUs) and Graphics Processing Units (GPUs), or customized hardware such as Field Programmable Gate Arrays (FPGAs) or Application-Specific Integrated Circuits (ASICs) [11]. While general-purpose compute engines continue

© The Author(s), under exclusive license to Springer Nature Switzerland AG 2024
G. Alsuhli et al., *Number Systems for Deep Neural Network Architectures*,
Synthesis Lectures on Engineering, Science, and Technology,
https://doi.org/10.1007/978-3-031-38133-1_1

1

to dominate DNN processing in academic settings, the industry places greater emphasis on deploying DNNs in resource-constrained edge devices, such as smartphones or wearable devices, which are commonly used for various practical applications [3]. Whether DNNs are run on GPUs or dedicated accelerators, speeding up and/or increasing DNN hardware efficiency without sacrificing their accuracy continues to be a demanding task. The literature includes a large number of works that have been dedicated to highlighting the directions that can be followed to reach these goals [3, 4, 12–15]. Some examples of these directions are DNN model compression [16], quantization [13], and DNN efficient processing [4, 12]. One of the directions that have a great impact on the performance of DNNs, but has not been comprehensively covered yet is the DNN number representation.

As the compute engines use a limited number of bits to represent values, real numbers cannot be infinitely represented. The mapping between a real number and the bits that represent it is called number representation or number system or data format [17]. Figure 1.1 shows an example to illustrate how a number can be represented differently with different number systems and how the choice of the number system directly affects the number of bits and cause different approximation to be happen. As a result, number representation has a great impact on the performance of both general-purpose and customized compute engines. DNNs encompass the learning of millions or even billions of parameters during model construction. As a result, the sheer volume of data associated with DNNs becomes substantial, requiring significant processing capabilities. Consequently, the choice of data representation format becomes crucial, impacting various aspects such as data precision, storage requirements, memory communication, and the implementation of arithmetic hard-

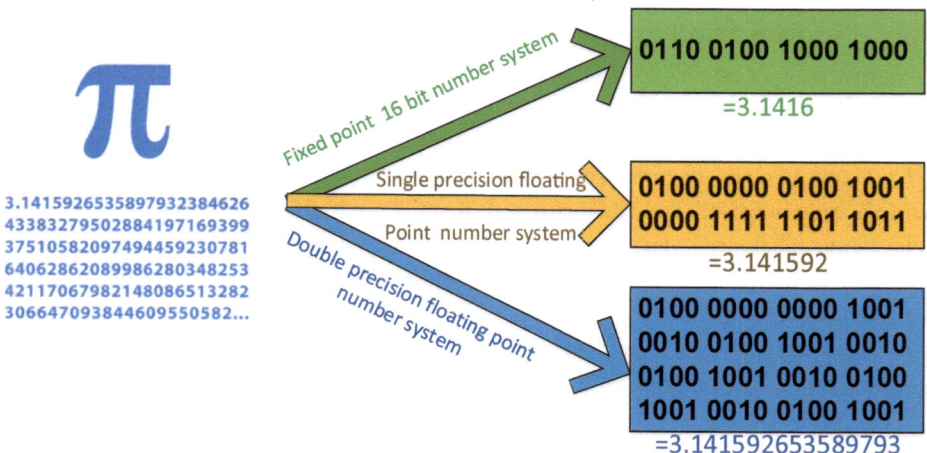

Fig. 1.1 An illustrative example of how the number $\pi = \frac{22}{7}$ can be represented using three well-known number systems: fixed point, single-precision floating point, and double-precision floating point. The approximations associated with each of these representations are illustrated

ware [18]. These factors, in turn, significantly influence key performance metrics of DNN architectures, including accuracy, power consumption, throughput, latency, and cost [12].

To this end, there is a significant body of literature that has focused on assessing the suitability of specific number systems for DNNs, modifying conventional number systems to fit DNN workloads, or proposing new number systems tailored for DNNs. Some of the leading companies, such as Google [19], NVIDIA [20], Microsoft [18], IBM [21], and Intel [22–24], have contributed in advancing the research in this field. A comprehensive discussion of these works will be helpful to furthering the research in this field.

While conventional number systems like Floating Point (FLP) and Fixed Point (FXP) representations are frequently used for DNN engines, several unconventional number systems are found to be more efficient for DNN implementation. Such alternative number systems are presented in this book and include the Logarithmic Number System (LNS), Residue Number System (RNS), Block Floating Point Number System (BFP), Dynamic Fixed Point Number System (DFXP), and Posit Number System (PNS). This book aims to provide a comprehensive discussion about alternative number systems for more efficient representations of DNN data. As an extension of our survey paper [25], it delves deeper into these alternative representations, offering an expanded discussion. The impact of these number systems on the performance and hardware design of DNNs is considered. In addition, this book highlights the challenges associated with each number system and various solutions that are proposed for addressing them. The reader will be able to understand the importance of an efficient number system for DNN, learn about the widely used number systems for DNN, understand the trade-offs between various number systems, and consider various design aspects that affect the impact of number systems on DNN performance. In addition, the recent trends and related research opportunities will be highlighted.

1.2   Book Organization

The structure of the book is summarized as follows.

- Chapter 2 provides a background of DNNs including their basic operations, types, main phases (training and inference), and an overview of their hardware implementations.
- Chapter 3 gives an overview of conventional number systems and their utilization for DNNs.
- Chapter 4 classifies the DNNs that adopt the logarithmic number system.
- Chapter 5 describes the concepts behind the residue number system and its employment for DNNs.
- Chapter 6 describes the block floating point representation and the efforts done to make it suitable for DNNs implementation.
- Chapter 7 discusses the dynamic fixed point format and the work done to calibrate the parameters associated with this format.

602619_1_En_1_Chapter-print ☑TYPESET ☐DISK ☐LE ☑CP Disp.:26/8/2023 Pages: 94 Layout: German_T5

- Chapter 8 explains various DNN architectures that utilize Posits and the advantages and disadvantages associated with these architectures.
- Chapter 9 concludes the book and provides insight into recent trends and research opportunities in the field of DNN number systems.

References

1. Deng, L., Li, J., Huang, J.T., Yao, K., Yu, D., Seide, F., Seltzer, M., Zweig, G., He, X., Williams, J., et al.: Recent advances in deep learning for speech research at Microsoft. In: IEEE International Conference on Acoustics, Speech and Signal Processing, pp. 8604–8608. IEEE (2013)
2. Krizhevsky, A., Sutskever, I., Hinton, G.E.: ImageNet classification with deep convolutional neural networks. Commun. ACM. **60**(6), 84–90 (2017)
3. Guo, Y.: A survey on methods and theories of quantized neural networks (2018). arXiv:1808.04752
4. Sze, V., Chen, Y.H., Yang, T.J., Emer, J.S.: Efficient processing of deep neural networks: a tutorial and survey. Proc. IEEE. **105**(12), 2295–2329 (2017)
5. Gupta, A., Anpalagan, A., Guan, L., Khwaja, A.S.: Deep learning for object detection and scene perception in self-driving cars: survey, challenges, and open issues. Array. **10**, 100057 (2021)
6. Shewalkar, A.: Performance evaluation of deep neural networks applied to speech recognition: RNN, LSTM and GRU. J. Artif. Intell. Soft Comput. Res. **9**(4), 235–245 (2019)
7. Buhrmester, V., Münch, D., Arens, M.: Analysis of explainers of black box deep neural networks for computer vision: a survey. Mach. Learn. Knowl. Extr. **3**(4), 966–989 (2021)
8. Otter, D.W., Medina, J.R., Kalita, J.K.: A survey of the usages of deep learning for natural language processing. IEEE Trans. Neural Netw. Learn. Syst. **32**(2), 604–624 (2020)
9. Pustokhina, I.V., Pustokhin, D.A., Gupta, D., Khanna, A., Shankar, K., Nguyen, G.N.: An effective training scheme for deep neural network in edge computing enabled internet of medical things (IoMT) systems. IEEE Access. **8**, 107112–107123 (2020)
10. Alam, M., Samad, M., Vidyaratne, L., Glandon, A., Iftekharuddin, K.: Survey on deep neural networks in speech and vision systems. Neurocomputing. **417**, 302–321 (2020)
11. LeCun, Y.: Deep learning hardware: past, present, and future. In: IEEE International Solid-State Circuits Conference-(ISSCC), pp. 12–19. IEEE (2019)
12. Sze, V., Chen, Y.H., Yang, T.J., Emer, J.S.: Efficient processing of deep neural networks. Synth. Lect. Comput. Arch. **15**(2), 1–341 (2020)
13. Gholami, A., Kim, S., Dong, Z., Yao, Z., Mahoney, M.W., Keutzer, K.: A survey of quantization methods for efficient neural network inference (2021). arXiv:2103.13630
14. Wu, C., Fresse, V., Suffran, B., Konik, H.: Accelerating DNNs from local to virtualized FPGA in the cloud: a survey of trends. J. Syst. Arch. **119**, 102257 (2021)
15. Ghimire, D., Kil, D., Kim, S.h.: A survey on efficient convolutional neural networks and hardware acceleration. Electron. **11**(6), 945 (2022)
16. Choudhary, T., Mishra, V., Goswami, A., Sarangapani, J.: A comprehensive survey on model compression and acceleration. Artif. Intell. Rev. **53**(7), 5113–5155 (2020)
17. Gohil, V., Walia, S., Mekie, J., Awasthi, M.: Fixed-posit: a floating-point representation for error-resilient applications. IEEE Trans. Circuits Syst. II Express Briefs. **68**(10), 3341–3345 (2021)
18. Darvish Rouhani, B., Lo, D., Zhao, R., Liu, M., Fowers, J., Ovtcharov, K., Vinogradsky, A., Massengill, S., Yang, L., Bittner, R., et al.: Pushing the limits of narrow precision inferencing at cloud scale with Microsoft floating point. Adv. Neural Inf. Process. Syst. **33**, 10271–10281 (2020)

602619_1_En_1_Chapter-print ☑ TYPESET ☐ DISK ☐ LE ☑ CP Disp.:26/8/2023 Pages: 94 Layout: German_T5

19. Wang, S., Kanwar, P.: BFloat16: The secret to high performance on cloud TPUs. Google Cloud Blog **30** (2019)
20. Choquette, J., Gandhi, W., Giroux, O., Stam, N., Krashinsky, R.: Nvidia A100 tensor core GPU: performance and innovation. IEEE Micro. **41**(2), 29–35 (2021)
21. Gupta, S., Agrawal, A., Gopalakrishnan, K., Narayanan, P.: Deep learning with limited numerical precision. In: International Conference on Machine Learning, pp. 1737–1746. PMLR (2015)
22. Kalamkar, D., Mudigere, D., Mellempudi, N., Das, D., Banerjee, K., Avancha, S., Vooturi, D.T., Jammalamadaka, N., Huang, J., Yuen, H., et al.: A study of BFLOAT16 for deep learning training (2019). arXiv:1905.12322
23. Köster, U., Webb, T., Wang, X., Nassar, M., Bansal, A.K., Constable, W., Elibol, O., Gray, S., Hall, S., Hornof, L., et al.: Flexpoint: An adaptive numerical format for efficient training of deep neural networks. Adv. Neural Inf. Process. Syst. **30** (2017)
24. Popescu, V., Nassar, M., Wang, X., Tumer, E., Webb, T.: Flexpoint: Predictive numerics for deep learning. In: 2018 IEEE 25th Symposium on Computer Arithmetic (ARITH), pp. 1–4. IEEE (2018)
25. Alsuhli, G., Sakellariou, V., Saleh, H., Al-Qutayri, M., Mohammad, B., Stouraitis, T.: Number systems for deep neural network architectures: a survey. arXiv preprint arXiv:2307.05035 (2023)

602619_1_En_1_Chapter-print ☑TYPESET ☐DISK ☐LE ☑CP Disp.:26/8/2023 Pages: **94** Layout: **German_T5**

Deep Neural Networks Overview

2

Abstract

The following Chapter introduces the reader to basic DNN concepts. It provides the definition of the basic DNN operations which are directly related to the underlying numbering system. It also decibels popular DNN models used in modern AI systems. Finally, an overview of the standard training procedure (gradient descent, backpropagation) is given.

2.1 DNN Evolution

The term "Neural Networks" originated from the efforts to mathematically model the information processing mechanism of biological systems. The McCulloch-Pitts model, developed in 1943, proposed a basic function for a neuron that involves applying a linear function to an input vector, followed by a non-linear decision or activation function that produces an output. By connecting multiple computing units (neurons) and organizing them into layers, the first forms of neural networks were created. Since then, neural network research has made tremendous progress, with simple and small networks evolving into complex architectures with multiple layers, forming the basis of deep learning. In Deep Neural Networks (DNNs), the extraction of the essential input features becomes part of the training process, resulting in impressive results in pattern recognition applications such as computer vision and natural language processing. The progress in neural network research can be attributed to advancements in network architectures, learning algorithms, and the availability of large training datasets. Additionally, improvements in hardware, including CPUs, GPUs, as well as specialized application-specific (ASIC) circuits, have also been a major factor in the success of

© The Author(s), under exclusive license to Springer Nature Switzerland AG 2024
G. Alsuhli et al., *Number Systems for Deep Neural Network Architectures*,
Synthesis Lectures on Engineering, Science, and Technology,
https://doi.org/10.1007/978-3-031-38133-1_2

602619_1_En_2_Chapter-print ☑ TYPESET ☐ DISK ☐ LE ☑ CP Disp.:26/8/2023 Pages: 94 Layout: German_T5

DNN models. The new hardware architecture paradigms led to the efficient processing of large amounts of data and facilitated the training of large models, resulting in remarkable improvements in the accuracy of neural networks.

2.2   Basic DNN Operation

The basic arithmetic operation that a DNN node performs is the dot product of a weight vector and an input vector:

$$y = F \left(\sum_{i=0}^{L} W_i X_i + b \right) \tag{2.1}$$

X_i, W_i and b denote the input and the parameters (weights and biases) of the DNN node, respectively. Then, a non-linear activation function F is applied to the intermediate dot product result, S, to give the final output Y of each neuron. The selection of activation functions is presented in Sect. 5.3.4. Figure 2.1 depicts such a node, where the output Y feeds the next layer nodes. For a practical DNN implementation the processing of the huge number of weights and biases, as well as the associated data transfers, become a bottleneck for DNN processing. DNN processing can refer either to *inference*, where a trained model makes prediction about a given task, and *training*, where the model learns from data points of a given dataset related to a certain task.

2.3   Popular Network Types

These elementary processing units (neurons) can be arranged in a variety of different structures, resulting in different neural network architectures. Some of the most popular DNN models include Convolutional Neural Networks (CNNs), Recurrent Neural Networks (RNNs), and more recently, Transformers.

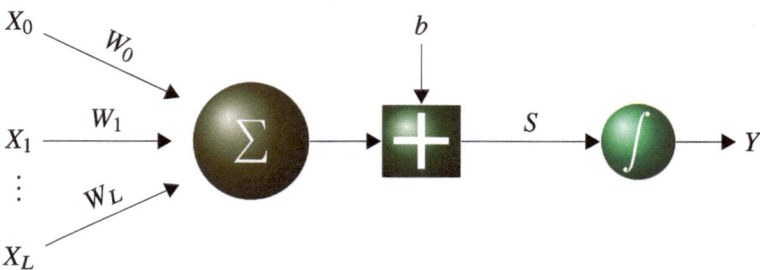

Fig. 2.1 A typical DNN node computes the dot product of its weight and input vectors and applies a non-linear activation function

602619_1_En_2_Chapter-print ☑ TYPESET ☐ DISK ☐ LE ☑ CP Disp.:26/8/2023 Pages: 94 Layout: German_T5

2.3.1 Convolutional Neural Networks

CNNs are a fundamental component of deep learning and have revolutionized many areas of machine learning, including computer vision and natural language processing. In CNNs, the network itself extracts, through the iterative adaptation of the filter coefficients, the input features with the highest information content. These features include edges, corners, and textures which can then be used to classify, segment, or recognize objects. An important property of convolution is its ability to preserve spatial information, enabling the use of CNNs for tasks such as object detection and localization as well.

CNNs can capture spatial dependencies inside the input. They are most commonly used in image and video processing tasks. Key aspects of CNNs are the sparse connections and the parameter sharing scheme [1]: Unlike traditional neural networks (Multi-layer Perceptron), that use dense matrix multiplication and each output unit interacts with each input unit, CNNs utilize smaller kernels with significantly less parameters (Fig. 2.2). In the case of a two-dimensional input image, for example, neighboring pixel regions are processed separately by neurons that share weights, i.e., transform the image in the same way. This cannot only dramatically decrease their computational complexity and memory requirements, but can also improve generalization and mitigate overfitting.

The elementary operation that each CNN neuron performs is the 2-d convolution. At each layer, a 3-d input, typically referred to as *feature map*, consisting of a number of channels is convolved with multiple *kernels* (filters). The convolution consists of a series of nested loops, where each kernel slides over the two spatial dimensions to calculate an output pixel output, according to Algorithm 1. C_{in} and C_{out} are the number of input and output channels, X, Y are the feature map's dimensions, F_X, F_Y are the filter's dimensions, I, O are the input and output feature maps, respectively, and W the weight tensor. Thus, the convolution simply consists of multiplying input feature maps with weights and accumulating these partial products. It follows that the basic arithmetic operation, in all CNNs, is the multiply-

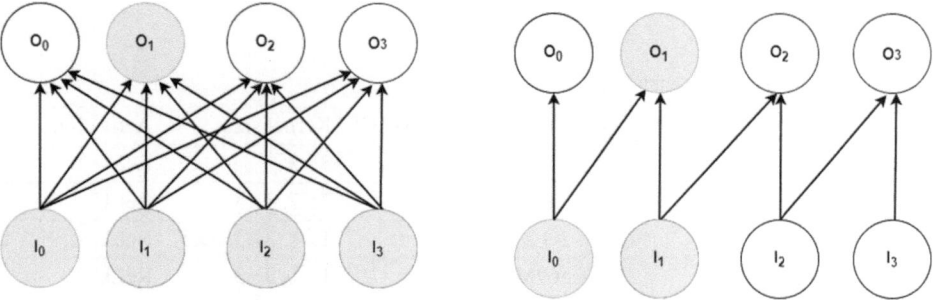

Fig. 2.2 Fully connected (left) vs convolutional network (right) connectivity. The weight-sharing scheme of the convolutional networks drastically reduces the number of parameters and computational complexity of the model and at the same time improves its generalization capability

602619_1_En_2_Chapter-print ☑ TYPESET ☐ DISK ☐ LE ☑ CP Disp.:26/8/2023 Pages: 94 Layout: German_T5

accumulate (MAC) operation. Multiple convolutional layers are stacked together to create deeper models, where each layer is assumed to represent information at a different level of abstraction. In the initial layers, filters of larger dimensions are used, while in the final layers, smaller ones are used.

Typically, pooling layers are added between certain convolutional layers. These layers calculate the mean or maximum of the input to gradually reduce its spatial dimensions while preserving their important features. The pooling operation also helps the obtained representation become approximately invariant to small translations of the input. A typical CNN structure is depicted in Fig. 2.3.

Popular CNN architectures that have given impassive results in computer vision tasks are presented in Table 2.1. These networks are composed of million parameters and require billions of operations, mainly MAC operations. They are also typically used as benchmarks to assess the performance of different hardware platforms

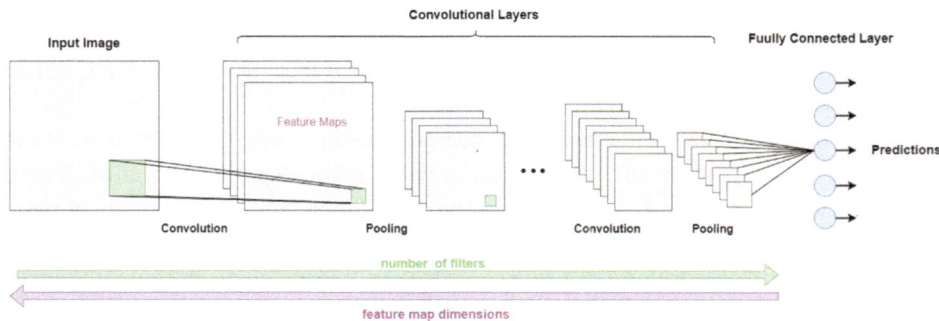

Fig. 2.3 Typical CNN structure. A series of convolutional and pooling layers is applied to the original image, to generate feature maps at different abstraction levels. The spatial dimensions of the feature maps is gradually decreased using pooling layers or convolution with stride greater than one, while the number of filters(kernels) increases when moving towards the end of the convolution pipeline. A fully connected layer, or generally a MLP classifier, is used at the end of the network to obtain the final predictions

Table 2.1 Popular CNN benchmarks

	# of parameters	# of operations (MAC)	Kernel sizes	Activation
VGG-19	143M	15.5G	3×3	ReLU
ResNet-50	25M	3.9G	7×7, 3×3	ReLU
Inception-v3	23M	2G	1×1, 3×3, 5×5	ReLU
MobileNet-v2	4M	569M	1×1, 3×3	ReLU
Yolo-v3	65M	18.7G	3×3	ReLU

602619_1_En_2_Chapter-print ☑TYPESET ☐DISK ☐LE ☑CP Disp :26/8/2023 Pages: 94 Layout: German_T5

Algorithm 1 Convolution loops

Require: $X = (x_1, x_2, \ldots, x_n)$
Ensure: the index i, such that $X \in \mathcal{I}_i$
1: **for** $c_o \leftarrow 0$ to C_{out} **do**
2:     **for** $x \leftarrow 0$ to X **do**
3:         **for** $y \leftarrow 0$ to Y **do**
4:             **for** $c_i \leftarrow 0$ to C_{in} **do**
5:                 **for** $f_x \leftarrow -F_X/2$ to $F_X/2$ **do**
6:                     **for** $f_y \leftarrow -F_Y/2$ to $F_Y/2$ **do**
7:                         $O[c_o][x][y] = I[c_i][x + f_x][y + f_y] \times W[c_i][f_x][f_y]$
8:                     **end for**
9:                 **end for**
10:             **end for**
11:         **end for**
12:     **end for**
13: **end for**

2.3.2 Recurrent Neural Networks

RNNs networks process information in successive time steps. The output of some neurons at a certain time can be fed back as input to other neurons at the next time step [2]. They, therefore, introduce a kind of memory in the processing of information and are particularly suitable when input samples exhibit temporal dependencies, such as in speech recognition or natural language processing applications. A typical RNN structure is depicted in Fig. 2.4. One of the most on RNN units is the Long Short-term Memory (LSTM cell), which is used to model to capture temporal dependencies between the input samples. Assume an input sequence $Y = \{y_1, y_2, \ldots, y_t\}$, where y_t is the input of the RNN at time t. An LSTM is defined by the following set of equations

$$i_t = \sigma\left(W^i x_t \oplus U^i h_{t-1} \oplus b^i\right), f_t = \sigma\left(W^f x_t \oplus U^f h_{t-1} \oplus b^f\right)$$
$$o_t = \sigma\left(W^o x_t \oplus U^o h_{t-1} \oplus b^o\right), \ c_t = f_t \odot c_{t-1} \oplus i_t \odot \tilde{c}_t,$$
$$\tilde{c}_t = \tanh\left(W^c x_t \oplus U^c h_{t-1} \oplus b^c\right), \ h_t = o_t \odot \tanh\left(c_t\right),$$

where W^k, U^k, and b^k, with $k = i, f, o, c$, are parameters of the RNN and are computed during the training process. Symbols $\odot$ and $\oplus$ denote element-wise multiplication and addition respectively. The input of the LSTM layer is x_t and for the input LSTM layer, it holds that $y_t = x_t$.

602619_1_En_2_Chapter-print ☑ TYPESET ☐ DISK ☐ LE ☑ CP Disp.:**26/8/2023** Pages: **94** Layout: **German_T5**

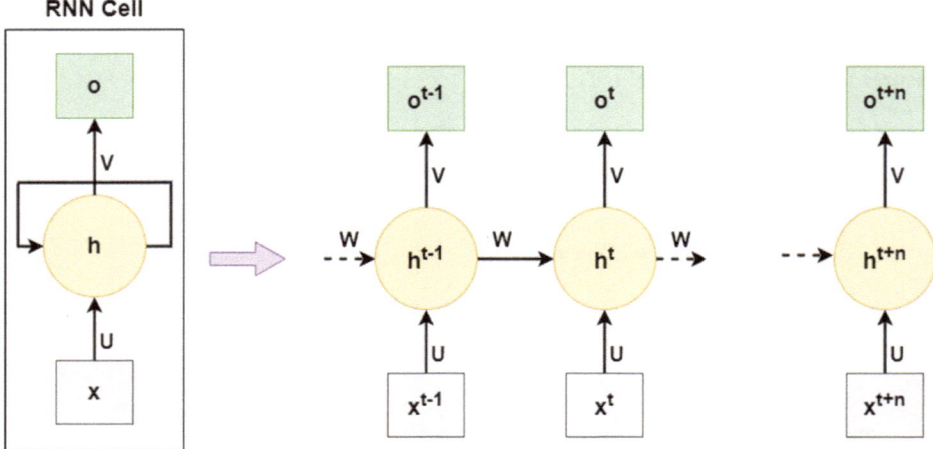

Fig. 2.4 Typical RNN structure. RNNs process sequential input and introduce a memory mechanism with the hidden state h of the cell, which is updated according to the previous state and current input (through trainable weight matrices W and U). The output of the cell is also a function of current state and input. Different RNN structures (like Gated Recurrent Unit or LSTM) are possible for the RNN cells

2.3.3 Transformers

Transformer neural networks are a type of deep learning model that have significantly advanced natural language processing (NLP) tasks. They were first introduced in 2017 by Vaswani et al. [3] and have since become the standard architecture for a broad range of NLP tasks, such as machine translation, question answering, and language modeling.

In contrast to traditional neural networks that process input data sequentially, transformers can simultaneously process all input tokens, thereby enabling more efficient computation. This is achieved through self-attention mechanisms that allow the model to weigh the importance of each token in the input sequence when computing the output. This mechanism allows capturing long-range dependencies in the input, which is crucial for NLP tasks where understanding the context of the text is essential.

The transformer model is composed of an encoder and a decoder. The encoder processes the input sequence and produces a representation of the input, while the decoder uses this representation to generate an output sequence. The transformer has demonstrated state-of-the-art performance on numerous NLP benchmarks, and has revolutionized the field of NLP.

602619_1_En_2_Chapter-print ☑ TYPESET ☐ DISK ☐ LE ☑ CP Disp.:26/8/2023 Pages: 94 Layout: German_T5

2.4   Activation Functions

Activation functions play a key role in NNs by introducing nonlinearity. This nonlinearity allows neural networks to generate more complex representations and approximate a wider variety of functions that would not be possible with a simple linear model. Common Activation functions utilized in modern deep-learning models include:

- *Sigmoid Function*

$$\sigma(x) = \frac{1}{1 + e^{-ax}} \tag{2.2}$$

The sigmoid function, also referred to as the logistic function, is one of the most commonly used activation functions and is particularly suitable for binary classification tasks. The sigmoid function maps its input to the range (0,1) and thus it can be directly interpreted as the class probability.

- *Hyperbolic Tangent Function*

$$\tanh(x) = \frac{e^{ax} - e^{-ax}}{e^{ax} + e^{-ax}} \tag{2.3}$$

Unlike the sigmoid function, which only produces positive values, the output of tanh lies between -1 and 1. It can sometimes result in faster training convergence. Sigmoid and tanh functions are also used in recurrent networks, like long short-term memory (LSTM) networks.

- *Rectified linear activation function (ReLU)*

$$\text{ReLU}(x) = \max(0, x) \tag{2.4}$$

The ReLU function is by far the most common choice in modern CNNs. It is a piecewise linear function that outputs the input if it is positive, otherwise, it will output zero. Unlike tanh and the sigmoid function, its output does not saturate. ReLU can overcome the vanishing gradient problem and it generally leads to faster convergence and better network performance. It is also more hardware friendly, thus it can further accelerate the training process.

- *Leaky ReLU*

$$\text{LeakyReLU}(x) = \max(ax, x) \tag{2.5}$$

where a is a small positive constant (typically 0.1 or 0.01) Leaky ReLU is a variation of ReLU which has a small slope for negative values, instead of mapping them to 0. It has been reported that networks using Leaky ReLU instead of simple ReLU can sometimes converge faster.

602619_1_En_2_Chapter-print ☑ TYPESET ☐ DISK ☐ LE ☐ CP Disp.:26/8/2023 Pages: 94 Layout: German_T5

- *Softmax function*

$$\sigma(\mathbf{X})_i = \frac{e^{x_i}}{\sum_{j=1}^{K} e^{x_j}} \text{ for } i = 1, \dots, K \tag{2.6}$$

The softmax function is most commonly used in the final layer of a multi-class classifier in order to normalize its outputs, converting them from weighted sum values into probabilities that sum to 1. Softmax applies the exponential function to each element x_i of the input vector $\mathbf{X}$ and normalizes these values by dividing by the sum of all these exponentials.

2.5   DNN Training

The goal of any training process is to adjust the parameters of the network, i.e., the neuron's weights so that the network approximates the desirable function based on the task definition. Therefore, once a network topology is determined, an error function E must be defined, which quantifies the deviation of the network output from the desired output for the set of input examples. Then, an appropriate algorithm is selected, which optimizes the network parameters with respect to the error function.

$$\frac{\partial E}{\partial w} = 0 \tag{2.7}$$

Since an analytical solution to this equation is usually not possible, iterative numerical methods are used. Generally, the methods for training a neural network can be classified into supervised, unsupervised, or competitive methods. The most commonly used training method is based on supervised learning through the gradient descent algorithm, in combination with the backpropagation method of error. Let C be a cost function of the minimum squares.

$$E = \frac{1}{N} \sum_{i=1}^{N} (Ytarg_i - Ypred_i)^2 \tag{2.8}$$

E represents the mean squared error for the set of the training vectors and is a measure of the distance between the network and the desired state. In the backpropagation training algorithm, the weight update is calculated based on the contribution of each weight to the total error. Considering the simplest form of ANN with two inputs and one output, the square of the error as a function of weights has the general form of a parabola with a hollow upward. The delta rule, also known as the method of gradient descent, follows the negative slope of the surface toward its minimum, moving the weight vector toward the ideal vector, according to the partial derivative of the error with respect to each weight.

$$Dw_{ij} = -a \frac{\partial E}{\partial w_{ij}} \tag{2.9}$$

602619_1_En_2_Chapter-print ☑ TYPESET ☐ DISK ☐ LE ☑ CP Disp.:26/8/2023 Pages: 94 Layout: German_T5

The weight updates are calculated iteratively using the chain derivation rule

$$Dw_{ij} = -a\frac{\partial E}{\partial y_i}\frac{\partial y_i}{\partial w_{ij}} \tag{2.10}$$

where y_i is the output of each neuron. By setting

$$d_i = -\frac{\partial E}{\partial y_i} = -\frac{\partial E}{\partial o_i}\frac{\partial o_i}{\partial y_i} \tag{2.11}$$

where o_i is the intermediate result of the neuron, before the application of the non-linear activation function.

For the output neurons, the coefficients d_i take the form

$$d_i = (b_i - o_i)\frac{\partial f(y_i)}{\partial y_i} \tag{2.12}$$

while for the hidden layer neurons, we have

$$d_i = \frac{\partial o_i}{\partial y_i}\sum_{k=1}^{K} d_k w_{ki} \tag{2.13}$$

The weight updates can be obtained as

$$Dw_{ij} = ad_i o_j \tag{2.14}$$

This method suffers from certain numerical and practical issues. The first problem is the large training time, mainly due to the non-optimal learning rate, which is different for each problem. The second problem is the saturation of the neuron's output, where during the training of the network, some weights are adjusted to large values and are not significantly modified afterward, causing the network to fall into a stagnant state. The third problem is the local minima, in which the network can become trapped. This happens because the error surface is not monotonic but consists of many "hills" and "valleys". Stochastic gradient descent and other variations of the above training methods based on the generalized Delta rule have been developed, overcoming many of these problems. One such method is the Adam method, which adjusts the damping of the learning rate via various stochastic methods as the training progresses.

References

1. Goodfellow, I., Bengio, Y., Courville, A.: Deep Learning. Springer (2015)
2. Bishop, C.: Pattern Recognition and Machine Learning. Springer (2006)
3. Vaswani, A., Shazeer, N., Parmar, N., Uszkoreit, J., Jones, L., Gomez, A.N., Kaiser, Ł., Polosukhin, I.: Attention is all you need. Adv. Neural Inf. Process. Syst. **30** (2017)

602619_1_En_2_Chapter-print ☑TYPESET ☐DISK ☐LE ☑CP Disp.:26/8/2023 Pages: 94 Layout: German_T5

Conventional Number Systems for DNN Architectures

3

Abstract

The two conventional number systems, namely the floating point and fixed point, are commonly used in almost all general-purpose DNN engines. While the FLP representation is usually used for modern computation platforms (e.g., CPUs and GPUs), where high precision is required, FXP is more common in low-cost computation platforms used for applications that require high speed, low power consumption, and small chip area. This chapter introduces these two representations and briefly discusses their utilization for implementing DNN hardware, in order to facilitate a comparison between conventional and unconventional number systems presented in subsequent chapters.

3.1 FLP for DNN Architectures

In the FLP number system, a number n is represented using a sign (1 bit), an exponent e (unsigned integer of length es), and a mantissa m (unsigned integer of length ms) (Fig. 3.1) and its value is given by

$$n = (-1)^s \times 2^{e-e_{max}} \times (1 + \frac{m}{2^{ms}}), \tag{3.1}$$

where $e_{max} = 2^{es-1} - 1$ is a bias used to ease the representation of both negative and positive exponents.

Although there are several FLP formats [1], the IEEE 754 FLP format [2], shown in Fig. 3.2, is the most common representation used by modern computing platforms [3, 4]. According to IEEE 754, the FLP can be of half, single, double, or quad-precision depending on the used bit-widths (e.g., for the single-precision FLP the bit-width is 32 bits and $es = 8$). The single-precision FLP, also called FLP32, is commonly used as a baseline to evaluate the

© The Author(s), under exclusive license to Springer Nature Switzerland AG 2024   17
G. Alsuhli et al., *Number Systems for Deep Neural Network Architectures*,
Synthesis Lectures on Engineering, Science, and Technology,
https://doi.org/10.1007/978-3-031-38133-1_3

602619_1_En_3_Chapter-print ☑ TYPESET ☐ DISK ☐ LE ☑ CP Disp.:26/8/2023 Pages: 94 Layout: German_T5

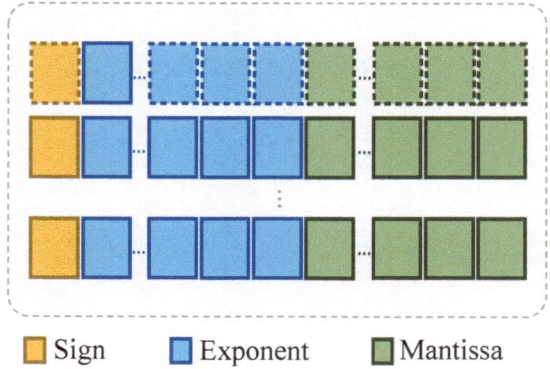

☐ Sign   ☐ Exponent   ☐ Mantissa

Fig. 3.1 The bit representation for the floating point number system

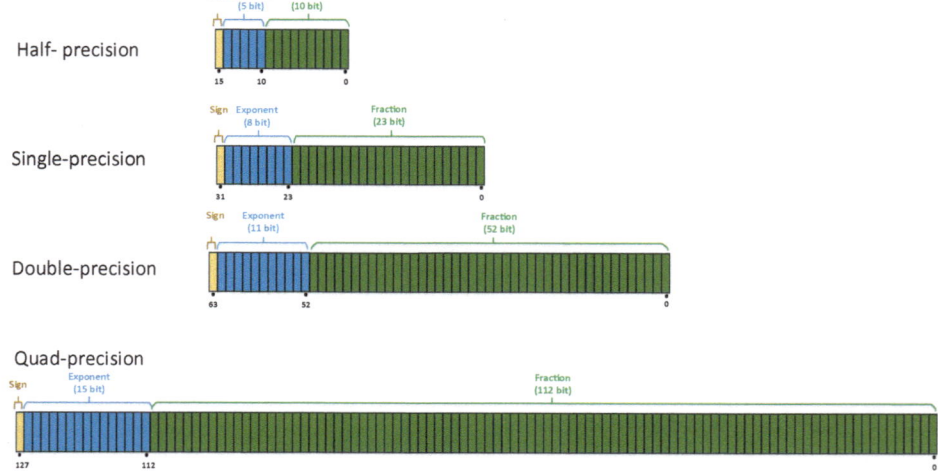

Fig. 3.2 IEEE 754 FLP formats

efficiency of other number representations. Unless otherwise stated, the performance degradation or enhancement is presented in comparison to the FLP32 format in this book as well.

3.1.1   Floating Point Arithmetic Operations

Floating point arithmetic operations include fundamental mathematical operations such as addition, subtraction, multiplication, and division. However, FLP arithmetic poses several challenges. Firstly, rounding errors can lead to a loss of precision. Secondly, FLP arithmetic incurs significant overhead compared to integer arithmetic due to the manipulation of mantissas and exponents to accurately compute results.

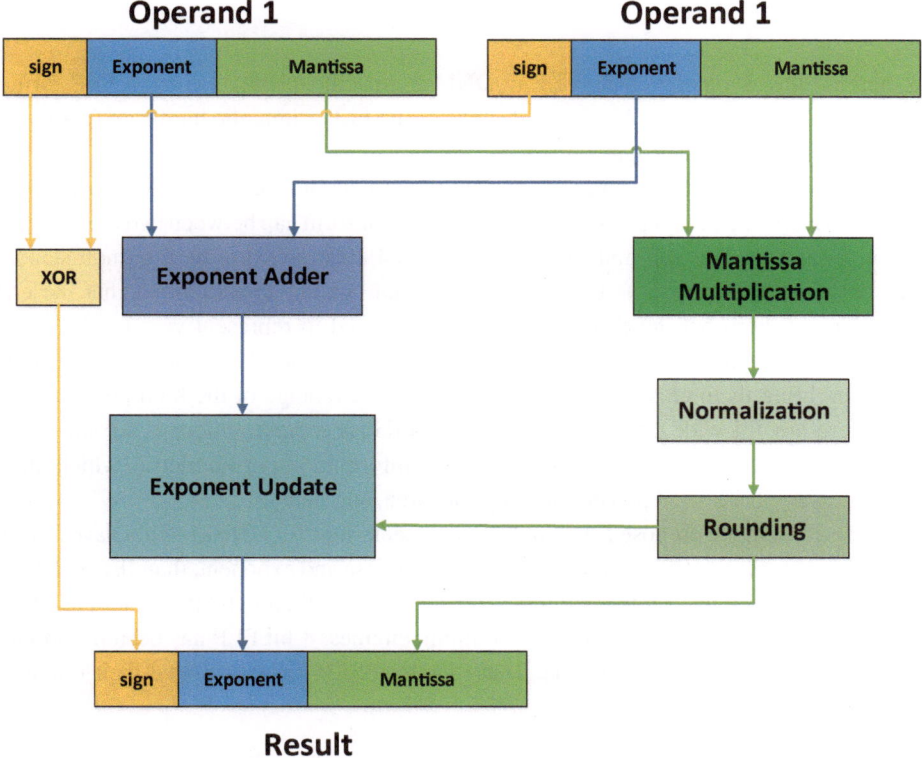

Fig. 3.3 Block diagram of an arithmetic unit dedicated to FLP multiplier

For instance, Fig. 3.3 illustrates the block diagram of the FLP multiplier architecture. It demonstrates that multiplying two FLP numbers involves adding their exponents, multiplying the mantissas, normalizing the resulting mantissa, and adjusting the exponent of the product [5]. Similarly, FLP addition requires comparing the operand exponents, shifting their mantissas (if the exponents differ), adding the mantissas, normalizing the sum mantissa, and adjusting the sum exponent [1].

The increased complexity of FLP32 arithmetic often necessitates the use of a dedicated unit called a Floating Point Unit (FPU) to perform FLP calculations [6]. However, the high power consumption and cost associated with the FPU limit its usage within embedded processing units like FPGAs [7]. As a result, the standard FLP32 format is rarely employed for building efficient DNN architectures [5].

602619_1_En_3_Chapter-print ☑ TYPESET ☐ DISK ☐ LE ☑ CP Disp.:**26/8/2023** Pages: **94** Layout: **German_T5**

3.1.2 FLP for DNNs

To increase the efficiency of the FLP in DNN architectures several custom FLP formats [8–11] have been proposed. Also, new designs of the FLP arithmetic hardware (mainly the multiplier) have been investigated [5, 12].

The standard FLP representations have a wide dynamic range,[1] as demonstrated by Table 3.1. However, these representations have a non-uniform gap between two representable numbers, resulting in a non-uniform error. Figure 3.4 illustrates how an 8-bit non-standard FLP representation represents the numbers between -15 to 15, and shows that the error is smaller near zero but increases when the FLP is used to represent very small or large numbers. As a result, for DNNs, quantization is typically used to scale the represented values and bring them closer to zero, thereby taking advantage of the high precision near zero. Therefore, the wide dynamic range of standard FLP representations is beyond what is usually required for DNNs [5], resulting in a low information-per-bit metric, which means an unnecessary increase in power consumption, area, and delay.

For this reason, the proposed custom FLP representations for DNNs mainly have reduced bit-width and a different allocation of the bits to mantissa and exponent, than IEEE 754. The bit-width is reduced to 19 bits in Nvidia's TensorFloat32 [9] and 16 bits in Google's Brain FLP (bfloat16) [8] formats used in DNN training engines. 8-bit FLP has been proposed to target the DNN inference in [10, 11]. These reduced FLP formats proved their efficiency

Table 3.1 A comparison between the smallest and largest numbers that can be represented using single and double precision FLP

	Smallest representable number	Largest representable number
Single precision FLP	1.18×10^{-38}	3.4×10^{38}
Double precision FLP	1.8×10^{-308}	2.23×10^{308}

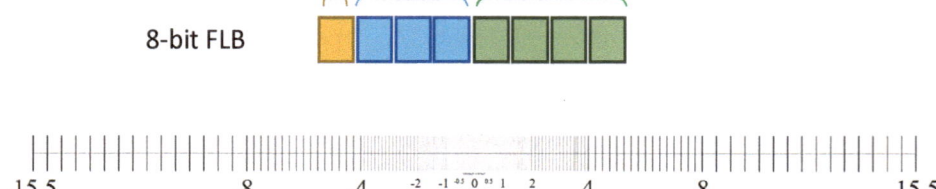

Fig. 3.4 Representable values of 8-bit non-standard FLP on number line in the range [-16, 16]. The Non-uniform gap between representable numbers is noticable

---

[1] The dynamic range of a number system is the ratio of the largest value that can be represented with this system to the smallest one.

602619_1_En_3_Chapter-print ☑ TYPESET ☐ DISK ☐ LE ☑ CP Disp.:26/8/2023 Pages: 94 Layout: German_T5

in replacing FLP32 with comparable accuracy, higher throughput, and smaller hardware footprint. It is worth noting that most of these custom FLP formats are used to represent data stored in memory (i.e., weights and activations), whereas, for internal calculations (e.g., accumulation and weight updates), FLP32 is used instead to avoid accuracy degradation [8, 11, 13].

In summary, the standard FLP representation has a massive dynamic range, which makes it a good choice for computationally intensive algorithms that include a wide range of values and require high precision. At the same time, the complex and power-hungry FLP calculations make FLP less attractive for DNN accelerators. This leads to using narrower custom FLP formats which require less hardware and memory footprint while preserving the performance of the standard FLP32. However, the utilization of the FLP format for DNN accelerators is relatively limited and it loses ground to a fixed point and other alternative representations.

3.2   FXP for DNN Architectures

The power inefficiency of the FLP arithmetic is the primary motivation to replace it with the FXP format for designing energy-constrained DNN accelerators. The bit representation for the fixed point number system is presented in Fig. 3.5. A real number n is represented in FXP with the sign, the integer, and the fraction parts. The fixed point format is usually indicated by $< I, F >$ where I and F correspond to the number of bits allocated to the integer and the fractional parts, respectively. In this book, we use the notations FXP8, for example, to denote the FXP representation with bit-width equal to 8, i.e., $I + F + 1 = 8$. FXP has a uniform gap between two representable numbers, equals to 2^{-F}, and, thus, a uniform error.

In FXP format, the separation between the integer and the fractional parts is implicit and usually done by specifying a scaling factor that is common for all data. Thus, the FXP

Fig. 3.5 The bit representation for several values using fixed point number system

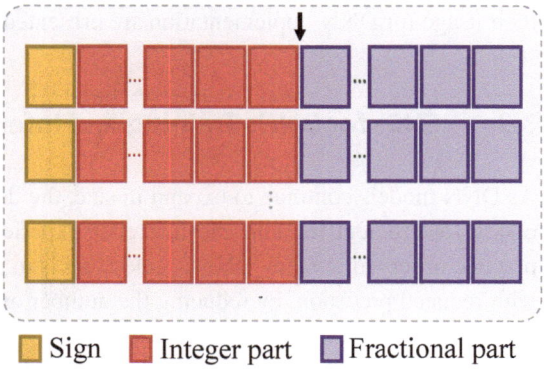

☐ Sign   ☐ Integer part   ☐ Fractional part

602619_1_En_3_Chapter-print ☑ TYPESET ☐ DISK ☐ LE ☑ CP Disp.:26/8/2023 Pages: 94 Layout: German_T5

number can be treated as an integer and, hence, integer arithmetic is used. Integer arithmetic requires substantially fewer logic gates to be implemented and it is faster and consumes much less chip area and power than FLP arithmetic. This makes FXP attractive to be used for DNN accelerators on edge. Moreover, the FXP allows for more reduction in the number of bits resulting in a significant reduction in the power consumption, storage requirements, and memory bandwidth [14].

On the other hand, the dynamic range of data represented by FXP is limited. For example, the smallest and largest representable numbers using FXP32$< 16, 16 >$, i.e. 16-bit for the integer part and 16-bit for the fraction part, are 1.53×10^{-5}, and 6.55×10^4, respectively. Compared to FLP32 in Table 3.1 with the same number of bits, FXP32 has a much smaller dynamic range. This makes FXP suitable to represent data with only a narrow range of values. Since this is not the case for most DNNs, using low-precision FXP for DNNs is challenging.

3.2.1 FXP for DNNs

Implementing hardware to accelerate the DNN inference is common in the literature and industry, some examples are in [15–20]. To enable utilizing low-precision FXP for DNNs, various approaches were adopted such as quantization[2] [21]. For instance, uniform quantization includes scaling weights and activations of DNN and mapping them to a restricted uniform range of values. These values can be represented by FXP with low-bit width. This allows lowering the number of bits to be less than 8 bits [18, 19, 22], and even as low as 2 bits (i.e., ternary DNNs [23–25]) or 1 bit (i.e., binary DNNs [26–30]). For more information about the FXP quantization, precision reduction, and binary DNNs the interested reader is referred to [3, 14, 21, 31].

In short, the FXP for DNN implementation offers great hardware efficiency at the expense of some accuracy degradation. Between the two extreme representations (FLP and FXP), there are several number systems that offer different trade-offs (Pareto optimal points) between the hardware efficiency and the acquired accuracy. These number systems and their usage for DNN implementation are presented in subsequent chapters of this book.

3.3   CNSs for DNNs Training and Inference

As DNN models continue to expand in size, the demand for computational resources necessary for efficient training and inference experiences a significant increase. A common practice observed in many DNN accelerators is to utilize the conventional number systems with reduced precision, by reducing the number of bits employed. This approach typically

---

[2] Quantization in general means to map continuous infinite values to a set of discrete finite values.

involves diminishing the precision of network parameters, inputs, and intermediate results to facilitate operations within reduced precision arithmetic units.

3.3.1 DNNs Inference Based on CNSs

The majority of AI accelerators, particularly those designed for low-power, high-throughput inference at the edge, utilize smaller word-length FXP arithmetic instead of FLP32 arithmetic, primarily due to the reasons mentioned earlier. For example, studies showed that performing addition on FXP8 consumes $3.3\times$ less energy than FXP32 addition and $30\times$ less energy than FLP32 addition. Similarly, when it comes to multiplication, the FXP8 multiplier consumes $15.5\times$ and $18.5\times$ less energy than an FXP32 and FP32 multipliers, respectively [32].

The most popular number systems used for DNN inference are the FXP8 and FXP16. 8-bit arithmetic generally results in faster systems with lower power consumption. More general-purpose architectures such as Nvidia's PASCAL GPUs and, Google's Tensor Processing Unit (TPU) also provide 8-bit instructions for DNN inference. 16-bit arithmetic is also very popular and is usually sufficient for many state-of-the-art CNN models for computer vision. For instance, the popular Eyeriss system [33] work with 16-bit integer arithmetic. The FXP32 is typically used in applications that require higher precision, such as natural language processing, where more complex DNN models are employed.

3.3.2 DNNs Training Based on CNSs

While DNN inference is relatively robust to precision reduction, lossless DNN training in reduced precision is a more challenging task. It has been shown that DNN inference can take place in FXP8 with a negligible accuracy drop for many state-of-the-art CNNs and computer vision benchmarks [32]. However, FXP arithmetic is found to be unsuitable for DNN training due to the greatly varying ranges of values that need to be handled during forward and backward passes. The computation of derivatives and backpropagation error terms require greater flexibility, since both very large and very small values need to be represented, thus FLP32 and more recently FLP16 arithmetic are still commonly used for DNN training.

References

1. Harris, D.M., Harris, S.L.: Hardware description languages. In: Digital Design and Computer Architecture, pp. 172–237 (2022)
2. IEEE: IEEE standard for floating-point arithmetic, IEEE Std 754-2019 (revision of IEEE 754-2008). Institute of Electrical and Electronics Engineers, New York (2019)

602619_1_En_3_Chapter-print ☑ TYPESET ☐ DISK ☐ LE ☑ CP Disp.:26/8/2023 Pages: 94 Layout: German_T5

3. Sze, V., Chen, Y.H., Yang, T.J., Emer, J.S.: Efficient processing of deep neural networks. Synthesis Lectures on Computer Architecture **15**(2), 1–341 (2020)
4. Courbariaux, M., Bengio, Y., David, J.P.: Training deep neural networks with low precision multiplications (2014). arXiv preprint arXiv:1412.7024
5. Leon, V., Paparouni, T., Petrongonas, E., Soudris, D., Pekmestzi, K.: Improving power of DSP and CNN hardware accelerators using approximate floating-point multipliers. ACM Trans. Embed. Comput. Syst. (TECS) **20**(5), 1–21 (2021)
6. Abdelaziz, H., Shin, J.H., Pedram, A., Hassoun, J., et al.: Rethinking floating point overheads for mixed precision DNN accelerators. Proc. Mach. Learn. Syst. **3**, 223–239 (2021)
7. Hassan, M.F., Hussein, K.F., Al-Musawi, B.: Design and implementation of fast floating point units for FPGAs. Indones. J. Electr. Eng. Comput. Sci. **19**(3), 1480–1489 (2020)
8. Wang, S., Kanwar, P.: BFloat16: the secret to high performance on cloud TPUs. Google Cloud Blog, vol. 30 (2019)
9. Choquette, J., Gandhi, W., Giroux, O., Stam, N., Krashinsky, R.: Nvidia A100 tensor core GPU: performance and innovation. IEEE Micro. **41**(2), 29–35 (2021)
10. Wu, C., Wang, M., Chu, X., Wang, K., He, L.: Low-precision floating-point arithmetic for high-performance FPGA-based CNN acceleration. ACM Trans. Reconfigurable Technol. Syst. (TRETS) **15**(1), 1–21 (2021)
11. Kang, H.J.: Short floating-point representation for convolutional neural network inference. IEICE Electronics Express, vol. 15, pp. 20180909 (2018)
12. Lee, H.J., Kim, C.H., Kim, S.W.: Design of floating-point MAC unit for computing DNN applications in PIM. In: International Conference on Electronics, Information, and Communication (ICEIC), pp. 1–7. IEEE (2020)
13. Narang, S., Diamos, G., Elsen, E., Micikevicius, P., Alben, J., Garcia, D., Ginsburg, B., Houston, M., Kuchaiev, O., Venkatesh, G., et al.: Mixed precision training. In: Proceeding 6th International Conference on Learning Representations (ICLR) (2018)
14. Sze, V., Chen, Y.H., Yang, T.J., Emer, J.S.: Efficient processing of deep neural networks: a tutorial and survey. Proc. IEEE. **105**(12), 2295–2329 (2017)
15. Kiyama, M., Amagasaki, M., Iida, M.: Deep learning framework with arbitrary numerical precision. In: 2019 IEEE 13th International Symposium on Embedded Multicore/Many-core Systems-on-Chip (MCSoC), pp. 81–86. IEEE (2019)
16. Wijeratne, S., Jayaweera, S., Dananjaya, M., Pasqual, A.: Reconfigurable co-processor architecture with limited numerical precision to accelerate deep convolutional neural networks. In: 2018 IEEE 29th International Conference on Application-specific Systems, Architectures and Processors (ASAP), pp. 1–7. IEEE (2018)
17. Suda, N., Chandra, V., Dasika, G., Mohanty, A., Ma, Y., Vrudhula, S., Seo, J.s., Cao, Y.: Throughput-optimized OpenCL-based FPGA accelerator for large-scale convolutional neural networks. In: Proceedings of the 2016 ACM/SIGDA International Symposium on Field-Programmable Gate Arrays, pp. 16–25 (2016)
18. Lo, C.Y., Sham, C.W.: Energy efficient fixed-point inference system of convolutional neural network. In: IEEE 63rd International Midwest Symposium on Circuits and Systems (MWSCAS), pp. 403–406. IEEE (2020)
19. Lin, D., Talathi, S., Annapureddy, S.: Fixed point quantization of deep convolutional networks. In: International Conference on Machine Learning, pp. 2849–2858. PMLR (2016)
20. Agrawal, A., Lee, S.K., Silberman, J., Ziegler, M., Kang, M., Venkataramani, S., Cao, N., Fleischer, B., Guillorn, M., Cohen, M., et al.: 9.1 a 7nm 4-core AI chip with 25.6 TFLOPS hybrid fp8 training, 102.4 TOPS INT4 inference and workload-aware throttling. In: 2021 IEEE International Solid-State Circuits Conference (ISSCC), vol. 64, pp. 144–146. IEEE (2021)

21. Gholami, A., Kim, S., Dong, Z., Yao, Z., Mahoney, M.W., Keutzer, K.: A survey of quantization methods for efficient neural network inference (2021). arXiv preprint arXiv:2103.13630

22. Anwar, S., Hwang, K., Sung, W.: Fixed point optimization of deep convolutional neural networks for object recognition. In: IEEE International Conference on Acoustics, Speech and Signal Processing (ICASSP), pp. 1131–1135 (2015). https://doi.org/10.1109/ICASSP.2015.7178146

23. Hwang, K., Sung, W.: Fixed-point feedforward deep neural network design using weights +1, 0, and -1. In: 2014 IEEE Workshop on Signal Processing Systems (SiPS), pp. 1–6. IEEE (2014)

24. Kim, S., Kim, H.: Zero-centered fixed-point quantization with iterative retraining for deep convolutional neural network-based object detectors. IEEE Access. **9**, 20828–20839 (2021)

25. Mellempudi, N., Kundu, A., Das, D., Mudigere, D., Kaul, B.: Mixed low-precision deep learning inference using dynamic fixed point. arXiv preprint arXiv:1701.08978 (2017)

26. Wang, Z., Lu, J., Tao, C., Zhou, J., Tian, Q.: Learning channel-wise interactions for binary convolutional neural networks. In: Proceedings of the IEEE/CVF Conference on Computer Vision and Pattern Recognition, pp. 568–577 (2019)

27. Rastegari, M., Ordonez, V., Redmon, J., Farhadi, A.: XNOR-Net: ImageNet classification using binary convolutional neural networks. In: European Conference on Computer Vision, pp. 525–542. Springer (2016)

28. Samragh, M., Hussain, S., Zhang, X., Huang, K., Koushanfar, F.: On the application of binary neural networks in oblivious inference. In: Proceedings of the IEEE/CVF Conference on Computer Vision and Pattern Recognition, pp. 4630–4639 (2021)

29. Yin, S., Jiang, Z., Seo, J.S., Seok, M.: XNOR-SRAM: In-memory computing SRAM macro for binary/ternary deep neural networks. IEEE J. Solid State Circuits **55**(6), 1733–1743 (2020)

30. Lin, Z., Courbariaux, M., Memisevic, R., Bengio, Y.: Neural networks with few multiplications (2015). arXiv preprint arXiv:1510.03009

31. Qin, H., Gong, R., Liu, X., Bai, X., Song, J., Sebe, N.: Binary neural networks: a survey. Pattern Recognit. **105**, 107281 (2020)

32. Sze, V., Chen, Y.H., Yang, T.J., Emer, J.S.: Efficient processing of deep neural networks: a tutorial and survey. Proc. IEEE **105**(12) (2017)

33. Chen, Y.H., Emer, J., Sze, V.: Eyeriss: A spatial architecture for energy-efficient dataflow for convolutional neural networks (2016). https://doi.org/10.1145/3007787.3001177

LNS for DNN Architectures

4

Abstract

This chapter discusses the Logarithmic number system (LNS). LNS is proposed as an alternative to conventional number representations to benefit from logarithm operation characteristics in implementing efficient hardware for costly arithmetic operations, such as multiplication which is massively required for DNNs. The chapter explains the proposals for LNS-based DNNs architectures and classifies them. Different classes are widely discussed and the challenges associated with each architecture are highlighted. The solutions presented so far for these challenges are spotlighted in this chapter as well.

4.1 Logarithmic Number System Overview

Proposals for LNS first emerged in the 1970s to implement the arithmetic operations of digital signal processing. The utilization of LNS for neural computing was first proposed in the late 90's [1]. Since then, using LNS to implement efficient hardware for DNN has become more popular. The main benefit of using LNS is in simplifying the implementation of the costly arithmetic operations required for DNN inference and/or training [2]. In addition, representing the data in LNS enables a reduction of the number of bits required to obtain the same DNN accuracy as with conventional number systems [3, 4]. In LNS, a real number n is represented with a logarithm of radix a of its absolute value ($\tilde{n} = \log_a(|n|)$) and a sign bit s_n[1] [2]. The number $\tilde{n}$ is represented using two's complement fixed point format [5], as shown in Fig. 4.1. The radix a of the logarithm is usually selected to be 2 for simpler hardware implementation. Throughout this chapter, we will use $a = 2$ as well.

---

[1] Some works use additional dedicated bit z_n to indicate when n equals zero [5, 6], while others use a special code to represent zero [2].

© The Author(s), under exclusive license to Springer Nature Switzerland AG 2024   27
G. Alsuhli et al., *Number Systems for Deep Neural Network Architectures*,
Synthesis Lectures on Engineering, Science, and Technology,
https://doi.org/10.1007/978-3-031-38133-1_4

Fig. 4.1 The bit representation for logarithmic number system

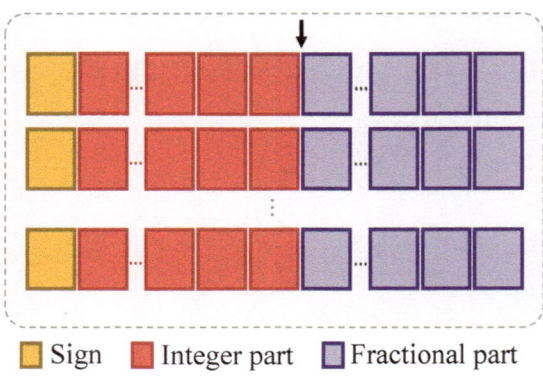

☐ Sign   ■ Integer part   ■ Fractional part

The main DNN operation that can be dramatically simplified using LNS is the multiplication by transforming it into linear (i.e., fixed point) addition. The LNS product $\tilde{p}$ of two real numbers n_1, and n_2 is calculated as follows

$$\tilde{p} = \tilde{n_1} \odot \tilde{n_2}, \tag{4.1}$$
$$= \log_2(|n_1| \times |n_2|),$$
$$= \tilde{n_1} + \tilde{n_2},$$

$$s_{\tilde{p}} = s_{n_1} X O R \, s_{n_2}, \tag{4.2}$$

where $\odot$ is the multiplication operation in LNS domain that can be implemented with a simple integer adder, and $s_{\tilde{p}}$ is the product sign, which is calculated by XORing the signs (s_{n_1} and s_{n_2}) of the two operands.

Existing proposals for LNS-based DNNs are for either using LNS for the whole DNN architecture from end-to-end, just for using the LNS-based multipliers, or for using logarithmic quantization for DNN weights and/or layer inputs. Based on this classification, LNS-based DNN architectures are discussed next by highlighting the challenges associated with each architecture and the solutions presented in the related work.

4.2   End-to-End LNS-Based DNN Architectures

End-to-end-LNS implementation utilizes the LNS for all blocks of the architecture, and thus, no conversion from or to conventional systems takes place. For this, the inputs (i.e., the dataset) and the weights[2] are assumed to be fed to DNN in LNS format. This task is usually performed offline and has no overhead on the implemented architecture. In this section, we review LNS-domain implementation of the main operations that are needed for DNN

---

[2] When the architecture targets DNN inference.

training and inference. The two types of DNNs that were implemented using LNS from end to end are CNNs [3, 4] and RNNs [6]. These two types of DNN have different architectures, but they share the same basic operations which are multiplication, addition, and activation functions. Since the multiplication operation becomes a linear addition in LNS-domain, the challenging part of this architecture is implementing LNS-addition and LNS-activation functions, which are discussed next.

4.2.1 Addition in LNS

As opposed to multiplication, performing addition in LNS is not straightforward. Let $\tilde{n}_1$ and $\tilde{n}_2$ be the two operands to be added in LNS. This LNS addition $\oplus$ is usually defined as follows

$$s\tilde{u}m = \tilde{n}_1 \oplus \tilde{n}_2, \tag{4.3}$$
$$= \log_2 |(-1)^{s_{n_1}} \times 2^{\tilde{n}_1} + (-1)^{s_{n_2}} \times 2^{\tilde{n}_2}|,$$

where $s\tilde{u}m$ is the LNS domain summation of the two operands, and s_{n_1} and s_{n_2} are their signs. As these operands can be negative or positive, $s\tilde{u}m$ is derived from (4.3) [4] such that

$$s\tilde{u}m = \begin{cases} \max(\tilde{n}_1, \tilde{n}_2) + \log_2(1 + 2^{-|\tilde{n}_1 - \tilde{n}_2|}), \\ \qquad \text{if } s_{n_1} = s_{n_2}, \\ \max(\tilde{n}_1, \tilde{n}_2) + \log_2(1 - 2^{-|\tilde{n}_1 - \tilde{n}_2|}), \\ \qquad \text{if } s_{n_1} \neq s_{n_2}, \end{cases} \tag{4.4}$$

$$s_{s\tilde{u}m} = \begin{cases} s_{n_1}, & n_1 > n_2, \\ s_{n_2}, & n_1 \leq n_2, \end{cases} \tag{4.5}$$

where $s_{s\tilde{u}m}$ is the sign of the summation. To reduce the computational complexity of calculating $s\tilde{u}m$, the term $\Delta_\pm = \log_2(1 \pm 2^{-|\tilde{n}_1 - \tilde{n}_2|})$ is approximated using look-up-tables (LUTs) [4, 6] or reduced to be implemented via bit-shifts [3, 4]. The LUT approximation requires using LUT of size r_{max}/r, where r_{max} is the range of values stored in the LUT and r is the resolution of the stored values. For the bit-shift implementation, the approximation in (4.6) is utilized to replace the calculation of $\log_2(1 \pm 2^{-|\tilde{n}_1 - \tilde{n}_2|})$ by a simple shift operation illustrated in (4.7).

$$\log_2(1 + x) = x, \text{ for } 0 < x < 1, \tag{4.6}$$

$$\log_2(1 \pm 2^{-|\tilde{n}_1 - \tilde{n}_2|}) = \pm \mathbf{BS}(1, -|\tilde{n}_1 - \tilde{n}_2|), \tag{4.7}$$

602619_1_En_4_Chapter-print ☑TYPESET ☐DISK ☐LE ☑CP Disp.:26/8/2023 Pages: 94 Layout: German_T5

where $\mathbf{BS}(b, d) = b \times 2^d$ means to shift the bits of b binary representation by $|d|$ Positions, to the left if d is negative and to the right otherwise. It has been shown in [6] that these two approximations are almost equivalent, However, LUTs require circuits with larger silicon and add extra delays to the system.

4.2.2 Activation Functions in LNS

Some activation functions can be transformed directly to the LNS domain by using the LNS operations to implement them. For example, the Leaky-ReLU for a number n in linear domain, shown in (4.8), is simply represented in LNS domain as in (4.9) [4].

$$LReLU(n|\alpha) = \begin{cases} \alpha n, \ n < 0, \\ n, \quad n \geq 0, \end{cases} \tag{4.8}$$

$$L\tilde{R}eLU((\tilde{n}, s_n)|\alpha) = \begin{cases} (\tilde{n} + \alpha, s_a = s_n), \ s_n = 1, \\ (\tilde{n}, s_a = s_n), \qquad s_n = 0, \end{cases} \tag{4.9}$$

where α is a constant, s_a is the sign after applying the activation, and $\tilde{n}$ and s_n are the logarithm and the sign of n. However, for more complicated activation functions, such as Sigmoid, tanh, and Softmax, more efficient hardware is obtained if these functions are approximated with piece-wise approximation that can be implemented using combinational logic [6]. The motivation of this is that the approximation becomes an additional source of non-linearity and places a low burden on the performance of the implemented DNN architecture. This is particularly the case if these functions are used within the training process which is inherently noisy [4]. As an example, Eq. (4.11) shows the LNS domain piece-wise approximation of tanh activation function in (4.10) [6].

$$\tanh(n) = \frac{1 - e^{-n}}{1 + e^{-n}} \tag{4.10}$$

$$\tilde{\tanh}(\tilde{n}, s_n) \approx \begin{cases} (0, s_a = s_n), \ \tilde{n} > 0, \\ (\tilde{n}, s_a = s_n), \ -10 < \tilde{n} \leq 0, \\ (0, s_a = 1), \ \tilde{n} < -10, \\ (0, s_a = 1), \ \tilde{n} = 0, z_n = 1, \end{cases} \tag{4.11}$$

602619_1_En_4_Chapter-print ☑TYPESET ☐DISK ☐LE ☑CP Disp.:26/8/2023 Pages: 94 Layout: German_T5

4.2.3   Summary and Discussion of End-to-End LNS-Based DNN Architectures

When LNS is used to represent DNN data from end to end, all operations needed to perform DNN training and / or inference must be implemented in the LNS domain. Multiplication is implemented using FXP addition, while other operations, such as addition and activation functions, need to be approximated. The presented approximation techniques that have been proposed for DNN implementation introduce insignificant loss in the performance of the implemented architectures. The classification accuracy degradation is found to be less than 1% for the studied end to end LNS-based CNN architectures [3, 4]. These work do not investigate the associated impact on the CNN hardware efficiency. However, an idea about the hardware efficiency is offered by LNS implementation of long short-term memory (LSTM) architecture where the area is saved by 36% for a 9-bit design [6], while the area savings decrease when the number of bits increased, due to the LUTs required for the LNS addition approximation.

4.3   LNS Multiplier-Based DNN Architectures

Since end-to-end LNS implementations of DNNs introduces complexity for implementing additions and activation functions, an alternative approach is to limit using the LNS to the multipliers. The focus here is to design an efficient LNS-based Multiplier that receives linear operands and produces a linear product as well.

4.3.1   LNS-Based Multiplier

To simplify the discussion and comparisons between various LNS multipliers proposed for DNN, some notations are introduced for the next sections. Let n be a Positive integer and its w-bit binary representation is $B_n = b_{w-1} \, b_{w-2} \ldots b_0$. Let b_k be the most significant '1' in B_n (k is called the characteristic number of n). The linear number n and its logarithm can be represented by

$$n = 2^k(1 + x), \tag{4.12}$$

$$\log_2(n) = k + \log_2(1 + x), \tag{4.13}$$

where $0 \leq x < 1$ is called the mantissa of n. Let $n_1 = 2^{k_1}(1 + x_1)$ and $n_2 = 2^{k_2}(1 + x_2)$ be the multiplier and multiplicand, respectively. The product of these operands and its logarithm are given by

$$n_1 \times n_2 = 2^{k_1 + k_2}(1 + x_1)(1 + x_2), \tag{4.14}$$

$$\log_2(n_1 \times n_2) = k_1 + k_2 + \log_2(1 + x_1) + \log_2(1 + x_2), \tag{4.15}$$

602619_1_En_4_Chapter-print ☑TYPESET ☐DISK ☐LE ☑CP Disp.:26/8/2023 Pages: 94 Layout: German_T5

The main idea of the logarithmic multiplier is to use a specific approximation based on the characteristics of the logarithms to simplify the product calculation by mainly using shift and add operations instead of hardware-intensive conventional multiplication. Given their effectiveness, many logarithmic multipliers have been proposed for image processing and neural computing [7–12]. Several of these multipliers were utilized to also build efficient DNN architectures. They are classified in this chapter into multipliers that use Mitchell's approximation, iterative logarithmic multipliers, double-sided error multipliers, and multipliers with explicit logarithm and antilogarithm modules.

4.3.2 Mitchell's Multiplier

According to Mitchell's algorithm [8], the logarithm of a number n is approximated with piece-wise straight lines as in (4.6). Thus, the logarithm of the product in (4.15) is approximated by the sum of the characteristic numbers and the mantissas of the operands as follows

$$\log_2(n_1 \times n_2) \approx k_1 + k_2 + x_1 + x_2. \tag{4.16}$$

The final product is obtained in (4.17) by applying the antilogarithm on (4.16) using the approximation in (4.6). Then, the product of two integers is calculated using add and shift operations, as

$$n_1 \times n_2 \approx \begin{cases} 2^{k_1+k_2}(1 + x_1 + x_2), & x_1 + x_2 < 1 \\ 2^{k_1+k_2+1}(x_1 + x_2), & x_1 + x_2 \geq 1 \end{cases} \tag{4.17}$$

Even though the error introduced by Mitchell's approximation is relatively high (up to 11% [9]), this multiplier showed no accuracy degradation for CNN architecture with 32- bit precision [13], while being 26.8% more power-efficient compared to conventional multipliers of the same number of bits.

To gain additional power efficiency over the one achieved by Mitchell's multiplier, a truncated-operand approach has been proposed [14]. Instead of using the whole operands, these operands are truncated and only their ω most significant bits are used to calculate the approximated product. For instance, selecting $\omega = 8$ allows for a more efficient multiplier that saves up to 88% and 56% of power when compared to an exact 32-bit FXP multiplier and a Mitchell's multiplier, respectively. The additional error introduced by this truncation caused an accuracy degradation of 0.2% for the ImageNet dataset. The significant power saving associated with the negligible performance degradation of this approach comes from the fact that the most significant part of the operand can be sufficient to provide an acceptable approximation [15, 16].

602619_1_En_4_Chapter-print ☑ TYPESET ☐ DISK ☐ LE ☑ CP Disp.:26/8/2023 Pages: 94 Layout: German_T5

4.3.3   Iterative Logarithmic Multiplier

The iterative logarithmic multiplier aims to reduce the error introduced by the approximation in (4.17) by adding correction terms. The calculation of these terms usually requires iterative multiplications that can be calculated in the same way as calculating the approximate product (see Fig. 4.2). These correction terms can be biased (always Positive) or unbiased (negative/Positive).

The product of two numbers in (4.14) can be written using biased correction terms [9] as

$$n_1 \times n_2 = P_{approx} + E, \tag{4.18}$$

where $P_{approx} = 2^{k_1+k_2}(1 + x_1 + x_2)$ is an approximate product that can be calculated using shift and add operations. $E = 2^{k_1+k_2} x_1 x_2$ is a correction term that is ignored in (4.17). Estimating the term E requires calculating the product $(2^{k_1} x_1)(2^{k_2} x_2) = (n_1 - 2^{k_1})(n_2 - 2^{k_2})$ iteratively, in the same way of calculating P_{approx}. Then,

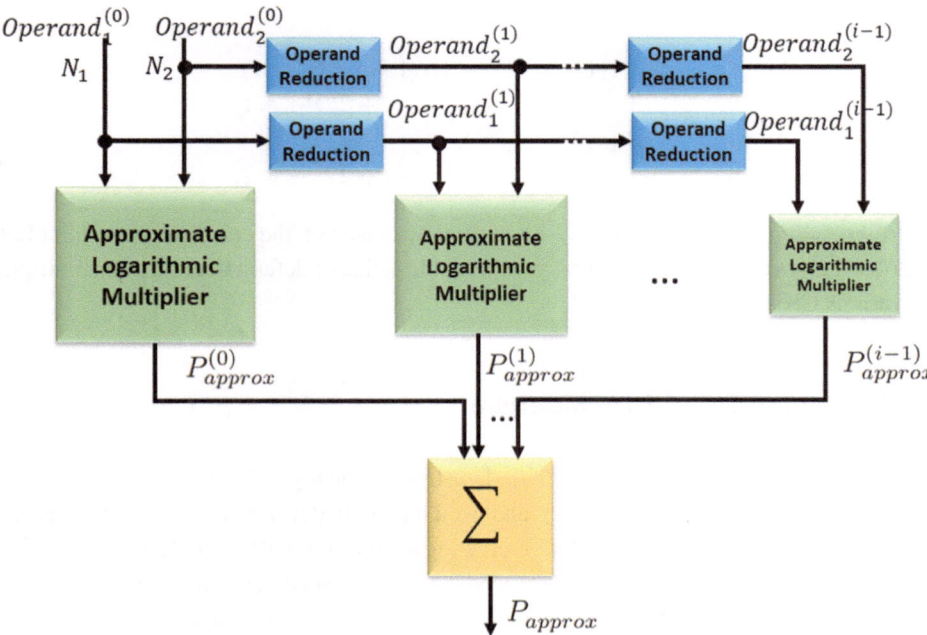

Fig. 4.2 Iterative logarithmic multiplier

602619_1_En_4_Chapter-print ☑TYPESET ☐DISK ☐LE ☑CP Disp.:26/8/2023 Pages: 94 Layout: German_T5

$$
\begin{aligned}
n_1 \times n_2 &= P^{(0)}_{approx} &&+ E^{(0)}, &&(4.19)\\
&= P^{(0)}_{approx} &&+ P^{(1)}_{approx} + E^{(1)},\\
&= P^{(0)}_{approx} &&+ P^{(1)}_{approx} + \cdots + P^{(i-1)}_{approx} + E^{(i-1)},
\end{aligned}
$$

where i is the number of iterations and $E^{(i-1)}$ is the error to be ignored after the ith iteration. Notice that when i equals the number of bits that have the value of '1' in the operands, then $E^{(i-1)} = 0$, and the exact product is produced. For each iteration, the new operands to be multiplied are obtained by removing the leading 'ones' from the original operands. For this reason, the correction terms can be calculated in parallel using one additional circuit for each iteration. Hence, there is a trade-off between the accuracy of the multiplication and the area and power overhead due to adding these correction circuits. For example, this iterative logarithmic multiplier with one iteration (i.e., one correction circuit) was able to save 10% on area and 20% on power consumption without any notable impact on the learning accuracy when it was used to implement the hardware of a relatively simple neural network and compared with the case of using floating point multiplier [17].

On the other hand, using unbiased iterative correction terms of (4.20) shows a better area and power reduction by up to 44.6% and 48.1%, respectively, compared to the multiplier designed with the error terms of (4.18) [18].

$$
E = \begin{cases}
((1 - x_1)2^{k_1} - 1)((1 - x_2)2^{k_2} - 1), \\
\qquad \text{if } x_1 + x_2 + 2^{-k_{1,2}} \geq 1 \\
(x_1 2^{k_1})(x_2 2^{k_2}), \\
\qquad \text{if } x_1 + x_2 + 2^{-k_{1,2}} < 1,
\end{cases}
\tag{4.20}
$$

where $k_{1,2} = max(k_1, k_2)$. Using 32-bit multipliers based on the correction terms in (4.20) and (4.18) gives classification accuracy comparable to that of default floating point multiplier for various CNN architectures [18].

4.3.4 Double-Sided Error Multiplier

The logarithmic approximation in (4.6) underestimates the logarithm value and results in an always negative error. Since tolerating and even having better performance in the presence of noise is an important feature of DNNs, creating a multiplier with "double-sided error" can enhance the implemented architecture of DNN in terms of accuracy and hardware efficiency [19]. To achieve this, another logarithmic approximation may be utilized [20].

In addition to the expression in (4.12), any integer number n can be represented as

$$
n = 2^{k+1}(1 - y), \tag{4.21}
$$

602619_1_En_4_Chapter-print ☑TYPESET ☐DISK ☐LE ☑CP Disp.:26/8/2023 Pages: 94 Layout: German_T5

where $0 \leq y < 1$. The two representations in (4.12) and (4.21) can be used to come up with a new logarithmic approximation with double-sided error [20], as

$$\log_2(n) \approx \begin{cases} k + x, & n - 2^k < 2^{k+1} - n. \\ k + 1 - y, & \text{otherwise.} \end{cases} \qquad (4.22)$$

To utilize this approximation, the two multiplied operands, n_1, and n_2, are transformed into the closest powers of two plus an additional negative or Positive term (a_1, a_2, respectively). Hence their product can be calculated as

$$n_1 \times n_2 = (2^{k_1} + a_1) \times (2^{k_2} + a_2), \qquad (4.23)$$
$$= 2^{k_1+k_2} + a_1 2^{k_2} + a_2 2^{k_1} + a_1 a_2.$$

The product is approximated to be the sum of the first three terms in (4.23), whereas the last term ($a_1 a_2$) is omitted as an approximation error. In fact, this approximation has a larger absolute error compared to Mitchell's approximation (of (4.6)). This can be observed as well from Fig. 4.3. However, the signed errors help with canceling the error and having higher classification accuracy by up to 1.4% compared to the case of using a conventional exact multiplier to implement CNNs [20]. This comes in addition to the better hardware efficiency indicated by 21.85% of power savings.

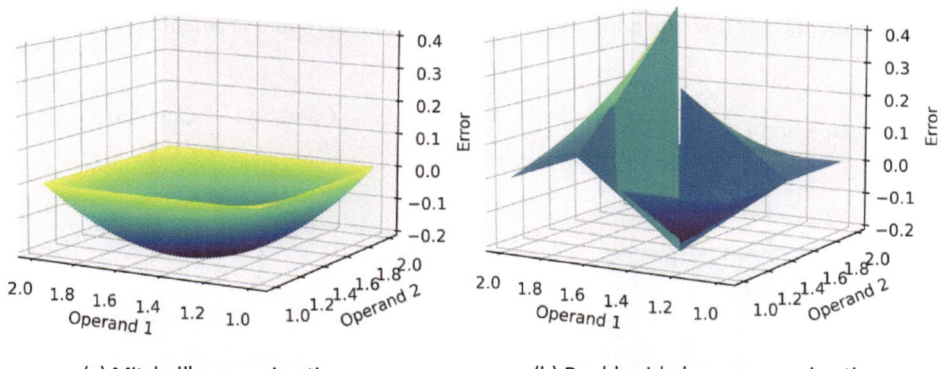

(a) Mitchell's approximation   (b) Double-sided error approximation

Fig. 4.3 An example compares the error arising from the utilization of Mitchell's and double-sided error Multipliers. The operands in this example lie within the range [0, 1], and the error is quantified as the difference between the approximated and exact values of the product

602619_1_En_4_Chapter-print ☑TYPESET ☐DISK ☐LE ☑CP Disp.:26/8/2023 Pages: 94 Layout: German_T5

4.3.5   Explicit Logarithm-Antilogarithm Multiplier

For the aforementioned multipliers (i.e., Mitchell's, iterative,..) there is no explicit module for logarithm and antilogarithm calculation. The implementation of these operations is not done explicitly, but their characteristics are used to transform the costly multiplication into simpler operations. On the other hand, the logarithmic multiplier can be designed by explicitly transforming the operands into the logarithmic domain, adding the operands, and finally returning back to the linear domain. The block diagram of this multiplier is presented in Fig. 4.4. As calculating the exact logarithm is very costly, the logarithm/antilogarithm operations are usually approximated using LUTs [21] or bit-level manipulation [22].

One approach that uses LUT-based approximated log/antilog multipliers for CNN is presented in [21]. Let n, represented in (4.12), be an operand in the linear domain. The logarithm of this operand is approximated as

$$\log_2(n) = k + \log_2(1 + x), \tag{4.24}$$
$$\approx k + Q_\beta(\log_2(Q_\gamma(1 + x))),$$

where Q_γ is the quantization used to represent $(1 + x)$ with γ bits in the linear domain, and β bits are guaranteed for the approximated logarithm in the log domain using Q_β. The mapping from $1 + x$ into $\log_2(1 + x)$ is obtained using a LUT of size $2^\gamma \beta$ bits. If the product in LNS domain is represented by $\tilde{p} = m.f$, where $f \in [0, 1)$ is the fraction part and m is the integer one, the approximated antilog of this product is calculated by

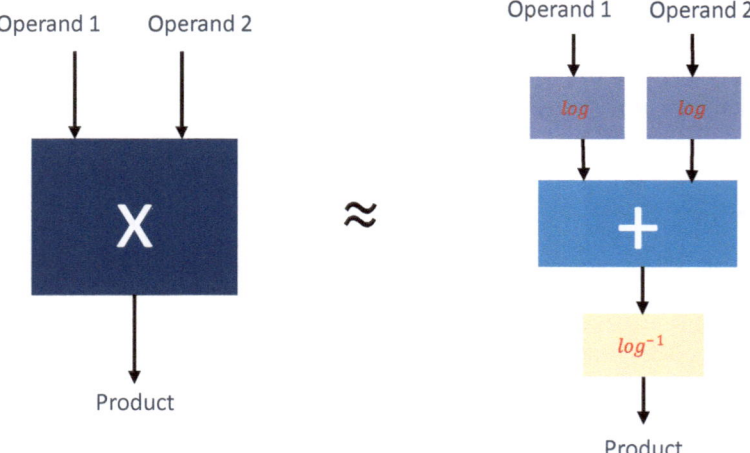

Fig. 4.4 Block diagram of explicit logarithm-antilogarithm multiplier

602619_1_En_4_Chapter-print ☑ TYPESET ☐ DISK ☐ LE ☑ CP Disp.:26/8/2023 Pages: 94 Layout: German_T5

$$\log_2^{-1}(\tilde{p}) = 2^m \, 2^f, \tag{4.25}$$
$$\approx 2^m \, (1 + Q_\alpha(2^f - 1)),$$

i.e., the term 2^m is implemented by a bit shift, whereas 2^f is approximated using a LUT. The LUT maps f to $Q_\alpha(2^f - 1)$, where Q_α is the applied quantization to limit the number of bits to α.

LUT-approximation is usually used to escape from the errors introduced by the approximation in (4.6). However, to keep the size of the required LUTs reasonable, the value of α, γ, and β should be kept as small as possible. This introduces a loss in accuracy. In addition, this approach is expected to be less hardware-friendly because of the needed overhead to implement these LUTs. Nevertheless, experimental results showed that integrating 16-bit LUT-based multiplier with a wide FXP accumulator results in a reduction in power consumption and area by up to 59% and 68%, respectively, in comparison to 16-bit FLP multiplier [21]. This comes in addition to achieving a negligible accuracy degradation ($<1\%$) for the CNN ResNet50 network trained on the ImageNet dataset.

Another approach for approximating log/antilog modules is using bit-level manipulation to innovate area and speed efficient logarithm or antilogarithm operations [22, 23]. Among these works, the two-region manipulation-based logarithm converter and bit correction-based antilogarithm converter [24] are used to implement an LNS multiplier that is exploited to build an efficient CNN accelerator design from area and delay point of view [22]. When this design is compared to conventional multiplier implementation, it saves up to 60% of the area-delay product. However, neither the accuracy of the CNN nor a comparison with other logarithmic multipliers has been reported for this design.

4.3.6 Summary and Discussion of LNS-Based Multipliers

LNS-based multipliers use the characteristics of the logarithm to transform the multiplication into simpler operations. Most of the proposed logarithmic multipliers for DNN architecture started from Mitchell's approximation to innovate their logarithmic multipliers. Table 4.1 compares various architectures that use LNS multipliers. We notice that these multipliers are used to implement efficient CNN architectures suitable for DNN inference rather than training. In addition, this table depicts that using the vanilla Mitchell's multiplier offers power-efficient hardware with comparable accuracy to the FLP32 multiplier of the same number of bits. However, reducing the number of bits requires a more accurate approximation with less average error than Mitchell's. When the LNS multiplier is designed with the characteristics of DNN in mind, such as the double-sided-error multiplier, the outcome is further reduction in the number of bits with significant power savings, while preserving and even enhancing the classification accuracy of the reported CNNs.

Table 4.1 Comparison of DNN architectures based on logarithmic multipliers. This table considers the targeted machine learning (ML) phase (inference/training), DNN type, the number system of the implemented arithmetic, the precision of the numbers indicated by number of bits, the dataset used to test the approach, and the tested DNN models. Comparison metrics (accuracy loss, power and area savings) are extracted from the related reference. Mostly, the enhancement or loss is reported in comparison to FLP32, when applicable. When the results is compared to another number system, this is indicated by mentioning this number system between Parentheses

References	ML Phase	DNN	DNN Arithmetic			Bits	Tested dataset	Tested model	Accuracy loss %	Power saving %	Area saving %
			Multiplier	Adder	Activation						
[13]	Inference	CNN	Mitchell's	Linear	Linear	32	MNIST, CIFAR-10	LeNet, Cudaonvnet	0	26.8	–
[14]	Inference	CNN	Truncated Mitchell's	Linear	Linear	32	MNIST, CIFAR-10, ImageNet	LeNet, cuda-convnet, AlexNet	0.2 (FXP)	96 (FXP)	81 (FXP)
[17]	Inference	NN	Iterative	Linear	Linear	18	Proben1	Custom	–	20	10
[18]	Inference	CNN	Iterative	Linear	Linear	32	CIFAR-10, ImageNet	NiN, AlexNet, GoogLeNet, ResNet-50	<0.5	–	–
[20]	Inference	DNN, CNN	Double-sided error	Linear	Linear	8	MNIST, CIFAR-10	Custom, Alexnet	-1.4[1]	21.85	–
[22]	Inference	CNN	Explicit Log/Antilog	Linear	Linear	32	–	Custom	–	–	~76
[21]	Inference	CNN	Explicit Log/Antilog	Linear	Linear	16	ImageNet	ResNet-50	<1	59 (FLP16)	68 (FLP16)

[1] Accuracy enhancement

602619_1_En_4_Chapter-print ☑ TYPESET ☐ DISK ☐ LE ☑ CP Disp.:**26/8/2023** Pages: **94** Layout: **German_T5**

4.4   Logarithmic Quantization for DNN Architectures

Logarithmic quantization involves representing a real number n with a sign and an integer exponent (integer power of two). The integer is usually an approximation of the logarithm $\log_2 |n|$ of the real number after applying the clipping and rounding [25, 26]. Logarithmic quantization has been employed in order to achieve efficient hardware implementation of CNNs [25, 26]. The main idea behind this is the that the multiplication by this integer exponent can be easily implemented in hardware by bit shifting. In CNNs, both the convolutional and fully-connected layers include matrix multiplication, i.e., the dot product between the weights W of each layer and the input activation X, which is the output of the previous layer after applying the non-linearity (e.g., ReLU). This matrix multiplication is usually performed using a number of multiply-and-accumulate operations when the conventional data representation is used to implement digital hardware, as shown in Fig. 4.5a. However, this dot product can be implemented more efficiently when Logarithmic quantization is utilized. Due to the non-uniform distribution of the weights and inputs, using nonuniform quantization, such as logarithmic quantization, is preferred over uniform quantization, such as when FXP is used [26].

Existing CNN architectures that use logarithmic quantization assume that weights and/or the inputs of the layer are quantized. When logarithmic quantization is applied to inputs only (i.e activations), Fig. 4.5b, or to weights only, Fig. 4.5c, the dot product becomes a simple bit shift operation followed by an accumulation. applying logarithmic quantization on the weights only shows insignificant accuracy degradation [27, 28] and significant power and area savings [29, 30], see Table 4.2. Quantizing the inputs only in LNS results in the same performance from an accuracy point of view [3, 31], however, with an additional linear-to-LNS module to be added. This module is responsible for transforming output activations to LNS before storing them in memory. This scheme has the advantage of requiring a smaller memory bandwidth as the stored activation is represented in LNS [3].

The works that apply logarithmic quantization to weights as well as to activations usually use a logarithm radix different from '2' [25, 26]. Then, the multiplication becomes an addition of the LNS quantized weights and activations followed by an approximation to decode this sum into the linear domain before implementing the accumulation. This add-decode-accumulate scheme adds a complication to hardware implementation, however, with comparable accuracy to the aforementioned logarithmic quantization schemes, as illustrated in Table 4.2.

602619_1_En_4_Chapter-print ✓ TYPESET ☐ DISK ☐ LE ✓ CP Disp.:26/8/2023 Pages: 94 Layout: German_T5

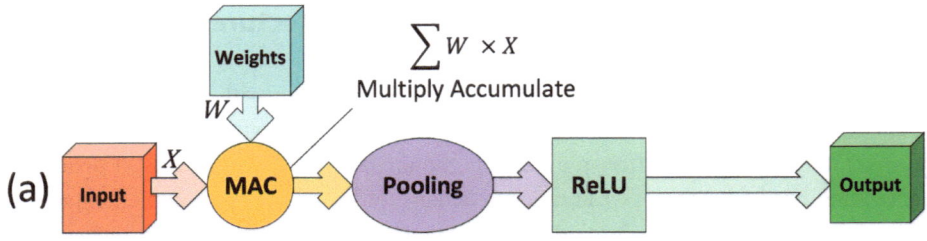

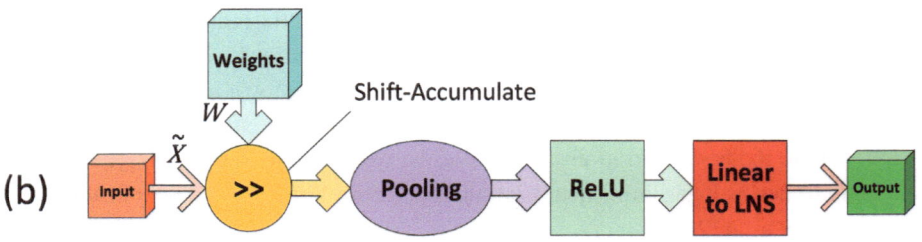

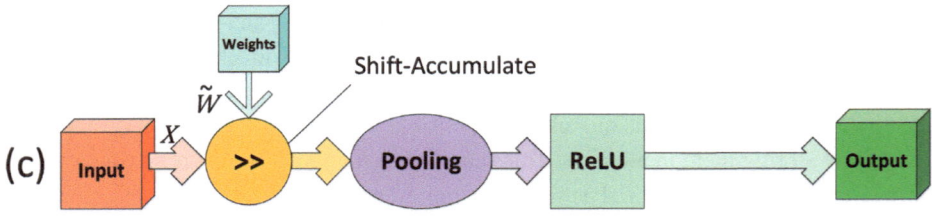

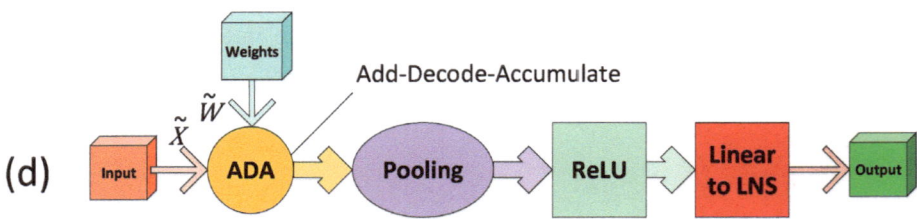

Fig. 4.5 Arithmetic processing elements of CNNs utilizing linear and various logarithmic quantization schemes **a** Linear quantization **b** Inputs logarithmic quantization **c** Weights logarithmic quantization **d** Inputs and weights logarithmic quantization

602619_1_En_4_Chapter-print ☑ TYPESET ☐ DISK ☐ LE ☑ CP Disp.:**26/8/2023** Pages: **94** Layout: **German_T5**

Table 4.2 Comparison of DNN architectures based on logarithmic quantization

References	ML phase	DNN	CNN arithmetic operations			Bits	Tested dataset	Tested model	Accuracy loss %	Power saving %	Area saving %
			Multiplier	Adder	Activation						
[30]	Inference	CNN	Shift	Linear	Linear	5	ImageNet	AlexNet, VGG16, ResNet34, DenseNet161	2.5	55.60 (FXP)	50.42 (FXP)
[29]	Inference	CNN	Shift	Linear	Linear	4	ImageNet	AlexNet	-3.9[1]	73.74 (FXP16)	58.03 (FXP16)
[27]	Inference	CNN	Shift	Linear	Linear	4 & 8	ImageNet	AlexNet, VGG16, ResNet-18/34, YOLOv2	<1	–	–
[28]	Inference	CNN	Shift	Linear	Linear	4	ImageNet	VGG16, AlexNet	<1	–	–
[31]	Training	DNN	Shift	Linear	Linear	7	MNIST	Custom	1.6	-90 (FXP32)	290 (FXP32)
[3]	Training	CNN	Shift	Linear	Linear	5	CIFAR10	AlexNet, VGG16	<1	–	–
[26]	Inference	CNN	Add-decode	Linear	Linear	8	CIFAR-10, ImageNet, Cityscapes	AllCNN, VGG16, Dilated	<3	22.3 (FXP)	35 (LUTs)
[25]	Training	CNN	Add-decode	Linear	Linear	8	CIFAR-10, ImageNet, SQuAD, GLUE	ResNet, BERT	<0.3	90	–

[1] Accuracy enhancement

602619_1_En_4_Chapter-print ☑ TYPESET ☐ DISK ☐ LE ☑ CP Disp.:**26/8/2023** Pages: **94** Layout: German_T5

References

1. Arnold, M.G., Bailey, T.A., Cupal, J.J., Winkel, M.D.: On the cost effectiveness of logarithmic arithmetic for backpropagation training on SIMD processors. In: Proceedings of International Conference on Neural Networks (ICNN'97), vol. 2, pp. 933–936. IEEE (1997)
2. Parhami, B.: Computing with logarithmic number system arithmetic: implementation methods and performance benefits. Comput. & Electr. Eng. **87**, 106800 (2020)
3. Miyashita, D., Lee, E.H., Murmann, B.: Convolutional neural networks using logarithmic data representation (2016). arXiv:1603.01025
4. Sanyal, A., Beerel, P.A., Chugg, K.M.: Neural network training with approximate logarithmic computations. In: IEEE International Conference on Acoustics, Speech and Signal Processing (ICASSP), pp. 3122–3126. IEEE (2020)
5. Alam, S.A., Garland, J., Gregg, D.: Low-precision logarithmic number systems: Beyond base-2. ACM Trans. Arch. Code Optim. (TACO). **18**(4), 1–25 (2021)
6. Kouretas, I., Paliouras, V.: Logarithmic number system for deep learning. In: International Conference on Modern Circuits and Systems Technologies (MOCAST), pp. 1–4. IEEE (2018)
7. Saadat, H., Bokhari, H., Parameswaran, S.: Minimally biased multipliers for approximate integer and floating-point multiplication. IEEE Trans. Comput. Aided Des. Integr. Circuits Syst. **37**(11), 2623–2635 (2018)
8. Mitchell, J.N.: Computer multiplication and division using binary logarithms. IRE Trans. Electron. Comput. **4**, 512–517 (1962)
9. Babić, Z., Avramović, A., Bulić, P.: An iterative logarithmic multiplier. Microprocess. Microsyst. **35**(1), 23–33 (2011)
10. Ansari, M.S., Cockburn, B.F., Han, J.: A hardware-efficient logarithmic multiplier with improved accuracy. In: Design, Automation & Test in Europe Conference & Exhibition (DATE), pp. 928–931. IEEE (2019)
11. Pilipović, R., Bulić, P.: On the design of logarithmic multiplier using radix-4 Booth encoding. IEEE access. **8**, 64578–64590 (2020)
12. Harsha, L., Jammu, B.R., Bodasingi, N., Veeramachaneni, S., SK, N.M.: A low error, hardware efficient logarithmic multiplier. Circuits Syst. Signal Process. **41**(1), 485–513 (2022)
13. Kim, M.S., Del Barrio, A.A., Hermida, R., Bagherzadeh, N.: Low-power implementation of Mitchell's approximate logarithmic multiplication for convolutional neural networks. In: Asia and South Pacific Design Automation Conference (ASP-DAC), pp. 617–622. IEEE (2018)
14. Kim, M.S., Del Barrio, A.A., Oliveira, L.T., Hermida, R., Bagherzadeh, N.: Efficient Mitchell's approximate log multipliers for convolutional neural networks. IEEE Trans. Comput. **68**(5), 660–675 (2018)
15. Sarwar, S.S., Venkataramani, S., Raghunathan, A., Roy, K.: Multiplier-less artificial neurons exploiting error resiliency for energy-efficient neural computing. In: Design, Automation & Test in Europe Conference & Exhibition (DATE), pp. 145–150. IEEE (2016)
16. Hashemi, S., Bahar, R.I., Reda, S.: DRUM: A dynamic range unbiased multiplier for approximate applications. In: IEEE/ACM International Conference on Computer-Aided Design (ICCAD), pp. 418–425. IEEE (2015)
17. Lotrič, U., Bulić, P.: Logarithmic multiplier in hardware implementation of neural networks. In: International Conference on Adaptive and Natural Computing Algorithms, pp. 158–168. Springer (2011)
18. Kim, H., Kim, M.S., Del Barrio, A.A., Bagherzadeh, N.: A cost-efficient iterative truncated logarithmic multiplication for convolutional neural networks. In: 2019 IEEE 26th Symposium on Computer Arithmetic (ARITH), pp. 108–111. IEEE (2019)

19. Ansari, M.S., Mrazek, V., Cockburn, B.F., Sekanina, L., Vasicek, Z., Han, J.: Improving the accuracy and hardware efficiency of neural networks using approximate multipliers. IEEE Trans. Very Large Scale Integr. (VLSI) Syst. **28**(2), 317–328 (2019)
20. Ansari, M.S., Cockburn, B.F., Han, J.: An improved logarithmic multiplier for energy-efficient neural computing. IEEE Trans. Comput. **70**(4), 614–625 (2020)
21. Johnson, J.: Rethinking floating point for deep learning (2018). arXiv:1811.01721
22. Juang, T.B., Lin, C.Y., Lin, G.Z.: Area-delay product efficient design for convolutional neural network circuits using logarithmic number systems. In: International SoC Design Conference (ISOCC), pp. 170–171. IEEE (2018)
23. Juang, T.B., Kuo, H.L., Jan, K.S.: Lower-error and area-efficient antilogarithmic converters with bit-correction schemes. J. Chin. Inst. Eng. **39**(1), 57–63 (2016)
24. Juang, T.B., Meher, P.K., Jan, K.S.: High-performance logarithmic converters using novel two-region bit-level manipulation schemes. In: Proceedings of 2011 International Symposium on VLSI Design, Automation and Test, pp. 1–4. IEEE (2011)
25. Zhao, J., Dai, S., Venkatesan, R., Liu, M.Y., Khailany, B., Dally, B., Anandkumar, A.: Low-precision training in logarithmic number system using multiplicative weight update (2021). arXiv:2106.13914
26. Vogel, S., Liang, M., Guntoro, A., Stechele, W., Ascheid, G.: Efficient hardware acceleration of CNNs using logarithmic data representation with arbitrary log-base. In: Proceedings of the International Conference on Computer-Aided Design, pp. 1–8 (2018)
27. Lu, T.Y., Chin, H.H., Wu, H.I., Tsay, R.S.: A very compact embedded CNN processor design based on logarithmic computing (2020). arXiv:2010.11686
28. Lee, E.H., Miyashita, D., Chai, E., Murmann, B., Wong, S.S.: LogNet: Energy-efficient neural networks using logarithmic computation. In: IEEE International Conference on Acoustics, Speech and Signal Processing (ICASSP), pp. 5900–5904. IEEE (2017)
29. Xu, J., Huan, Y., Zheng, L.R., Zou, Z.: A low-power arithmetic element for multi-base logarithmic computation on deep neural networks. In: IEEE International System-on-Chip Conference (SOCC), pp. 43–48. IEEE (2018)
30. Xu, J., Huan, Y., Jin, Y., Chu, H., Zheng, L.R., Zou, Z.: Base-reconfigurable segmented logarithmic quantization and hardware design for deep neural networks. J. Signal Process. Syst. **92**(11), 1263–1276 (2020)
31. Ueki, T., Iwai, K., Matsubara, T., Kurokawa, T.: Learning accelerator of deep neural networks with logarithmic quantization. In: 2018 7th International Congress on Advanced Applied Informatics (IIAI-AAI), pp. 634–638. IEEE (2018)

RNS for DNN Architectures

5

Abstract

Modern Deep Learning models keep growing in depth and number of parameters and require a huge amount of elementary arithmetic operations, the majority of which are multiply-add operations. These operations can be very efficiently implemented in the Residue Numbering System: RNS encoding allows for carry-free computations among the different residue channels, with inherent parallelism at the digit processing level. Arithmetic circuits for addition and multiplication become smaller and can operate on higher frequencies and with lower power consumption. In this Chapter, the basic RNS arithmetic operations and their hardware implementation are described. Moreover, RNS-based DNN architectures reported in the literature are presented and compared.

5.1   RNS for DNN Architectures

The number representation scheme utilized in realizing DNN architectures directly impacts the accuracy, speed, area, and energy dissipation. The Residue Number System can be an attractive choice for DNN accelerators due to its arithmetic properties. In this Chapter, a brief overview of the RNS is given, and several RNS-based architectures for AI applications reported in the literature are presented. Architectures are classified to *partially RNS-based*, where intermediate conversions to conventional representations between successive layers are used, and *end-to-end* RNS-based architectures, where the entire processing takes place in the RNS domain. The typical computation flow of these two types of systems is shown in Fig. 5.1.

In the Residue Number System, each number is represented as a tuple of residues with respect to a modulus set $\{m_1, m_2, \ldots, m_n\}$, which is called the *base* of the representation (Fig. 5.2). The dynamic range of the representation is given by

© The Author(s), under exclusive license to Springer Nature Switzerland AG 2024   45
G. Alsuhli et al., *Number Systems for Deep Neural Network Architectures*,
Synthesis Lectures on Engineering, Science, and Technology,
https://doi.org/10.1007/978-3-031-38133-1_5

602619_1_En_5_Chapter-print ☑ TYPESET ☐ DISK ☐ LE ☑ CP Disp.:26/8/2023 Pages: 94 Layout: German_T5

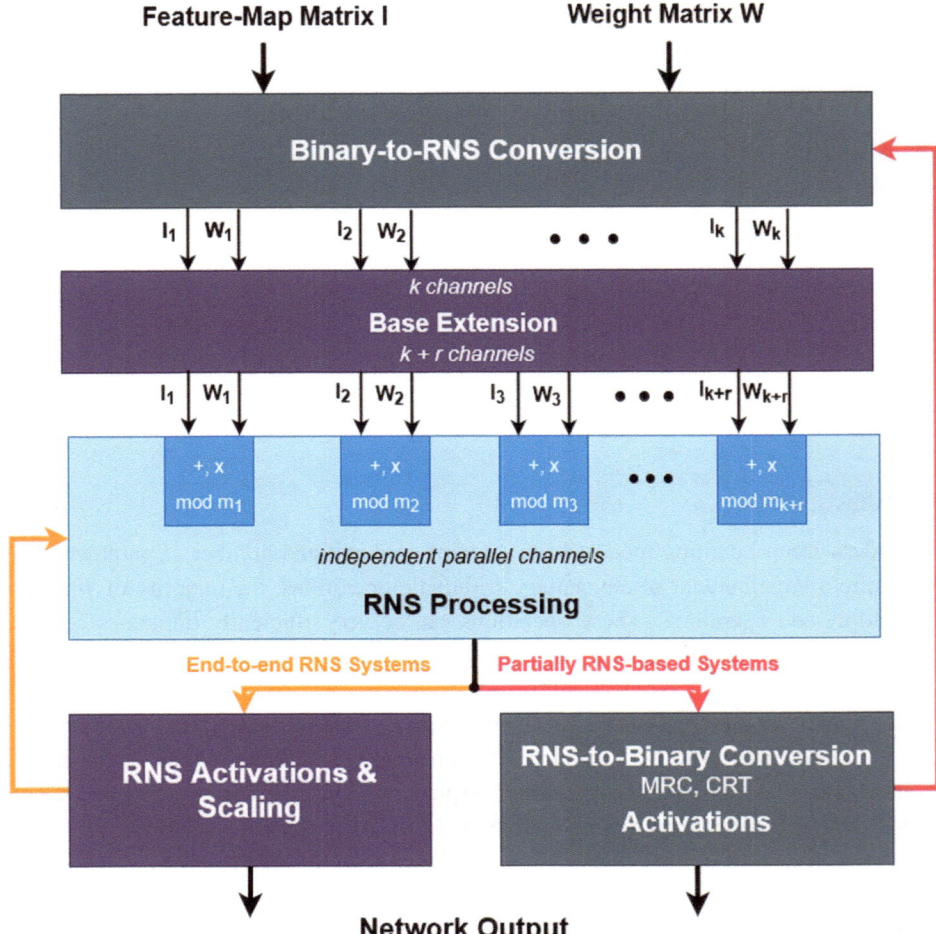

Fig. 5.1 Computation flow of RNS accelerators. Partially RNS-based accelerators utilize binary converters between successive layers for non-trivial RNS operations, while end-to-end systems perform all NN operations, including activation functions, in the RNS domain. Base-extension is usually required to increase the dynamic range before the accumulation of the partial products

$$R = \prod_{i=1}^{N} m_i. \tag{5.1}$$

If the moduli are *co-prime*, i.e.,

$$\gcd_{\substack{1 \le i,j \le N \\ i \ne j}} (m_i, m_j) = 1, \tag{5.2}$$

Fig. 5.2 RNS representation. Each integer is mapped to a set of residue channels. Addition and multiplication can be carried out independently and in parallel among these channels, reducing the critical path of the arithmetic circuits

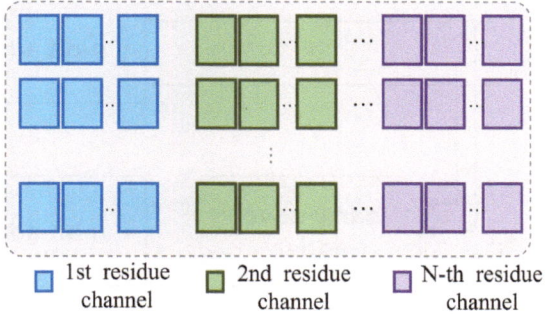

where gcd(·) denotes the greatest common divisor operation, each integer inside the range $[0, R)$ has a unique RNS representation

$$X \mapsto (x_1, x_2, \ldots, x_n), \ x_i = \langle X \rangle_{m_i}, \tag{5.3}$$

where $\langle \cdot \rangle_m$ is the modulo-m operator. Inverse transformation is generally harder and can be realized by means of the *Mixed Radix Conversion* or the *Chinese Remainder Theorem* [1].

5.2   RNS Addition and Multiplication

Due to the properties of the modulo operation, addition and multiplication can be done independently and in parallel for each residue channel, i.e., without inter-channel propagation of information. Suppose $A = (a_1, a_2, \ldots, a_n)$ and $B = (b_1, b_2, \ldots, b_n)$, then

$$a \oplus b = (\langle a_1 \oplus b_1 \rangle_{m_1}, \langle a_2 \oplus b_2 \rangle_{m_2}, \ldots, \langle a_n \oplus b_n \rangle_{m_n}), \tag{5.4}$$

where $\oplus$ can be either the addition or the multiplication operator. This property is what makes RNS very efficient in applications that require a large number of these operations, such as DSP applications and, more recently, Neural Network inference. This is because, by decomposing the computations into independent channels, long carry propagation chains are eliminated, thus arithmetic circuits can operate at higher frequencies, or with reduced power dissipation.

The general architecture of a modulo adder is shown in Fig. 5.3 [2]. The design consists of an n-bit, adder, where n is the size of the channel, that performs the addition of two numbers $a + b$, and a CSA adder which performs the computation of $a + b - m_i$ (modulo operation). The sign of the CSA result is used to select the correct result of the two adders.

The selection of moduli can significantly simplify the design of modulo arithmetic circuits. In case of moduli of the form 2^k the modulo operation translates into just keeping the k least significant bits, whereas in the case of $2^k - 1$, the output carry of the addition simply needs to be added to the result. In this case, end-around-carry adders can be used.

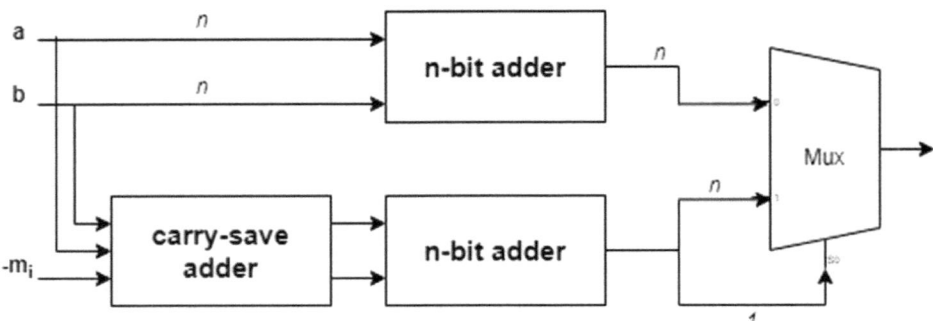

Fig. 5.3 Modulo adder

For channels of the form $2^k + 1$, diminished-1 arithmetic can be used [3], which basically involves an inverted end-around logic. If the size of the channel is large, then fast adder designs such as prefix adders must be utilized within each channel.

Modulo multiplication is a trickier operation, however the benefits of RNS can be greater. This is because of the (approximately) quadratic scaling of a multiplier with the input size. This means that, by decomposing a large multiplication into smaller ones, the energy and delay savings can be significant, providing that the overhead of the modulo is diminished. One approach for RNS multiplication is to perform regular multiplication of the two n-bit numbers and then use a reduction circuit to obtain the final result modulo m_i. This approach introduces, however, considerable overhead to the design, as the reduction of a $2n$-bit number to a n-bit number modulo m_i is not as straightforward as in the case of addition. A low complexity adder-based combinatorial multiplier has been proposed in [4], where the number of FAs required is minimized. Other multiplication techniques are based on intermediate RNS transformations, such as *core functions* [5] and *isomorphisms* [2], which are transformations that convert multiplication into addition. These transformations utilize look-up tables to convert RNS to an intermediate representation where multiplication is translated into addition.

In the case of modulo 2^k multiplication, regular multipliers operating only on the k LSBs can be used, whereas in the case of modulo-$(2^k - 1)$ diminished-1 arithmetic can be applied [6]. A end-around-carry multiplier which can be used for $2^k - 1$ channels is shown in Fig. 5.4. Due to the properties of the particular channel, the modulo operation is translated into simple bit re-ordering, thus no overhead is introduced.

Based on the above, most of the RNS designs reported in literature utilize the low-cost forms of moduli, which allow to fully exploit the RNS benefits (elimination of long carry chains) with minimal hardware overhead.

602619_1_En_5_Chapter-print ☑ TYPESET ☐ DISK ☐ LE ☑ CP Disp.:26/8/2023 Pages: 94 Layout: German_T5

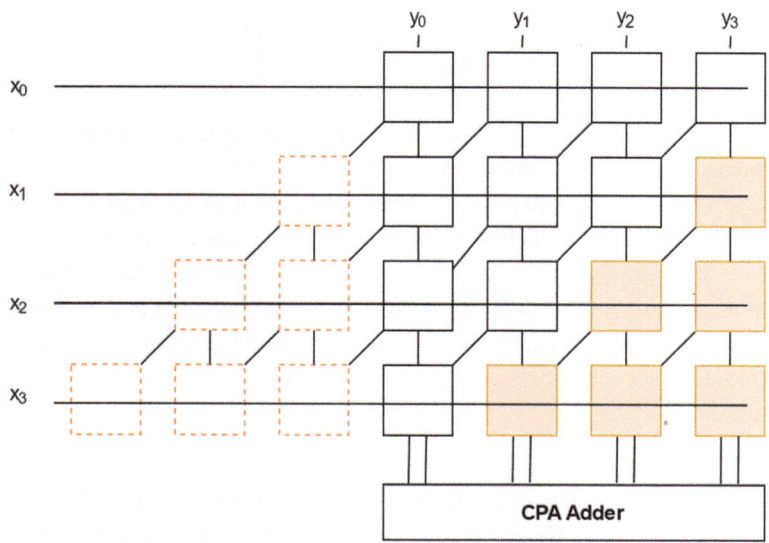

Fig. 5.4 Array multiplier for calculating $x \cdot y$ mod 15. At each level i, the full adders corresponding to the i most significant positions (dashed squares) are moved to the i least significant positions. This is possible because 2^{n+k} mod $(2^n - 1) = 2^k$. A modulo-15 carry-propagation adder is used to obtain the final result

5.3 Conversions and Non-trivial Operations

While addition and multiplication are very efficiently implemented in RNS, other operations such as sign detection, comparison and division, or the realization of non-linear activation functions are not straightforward to implement, as the require the combination of the RNS channels. A common approach is to use RNS-to-binary converters and then perform the operation in the binary domain.

5.3.1 RNS-to-Binary and Binary-to-RNS Conversions

Conversion to and from an RNS representation is a crucial for the performance of any RNS-based processing system. Especially for the architectures that perform frequent intermediate conversions (partially RNS-based) the overhead can be significant. The complexity of these converters largely depends on the particular base selection, namely the size, number and format of the moduli. While *Binary-to-RNS* or *forward* converters can have a relative simple hardware realization, following Eq. 5.3, especially if particular forms of moduli are used, *RNS-to-binary* or *inverse* converters are generally harder to implement. Extensive bibliography exists for this topic. The most commonly used approaches are the Chinese Remainder Theorem (CRT) and the Mixed Radix Conversion (MRC) [7]. The CRT is expressed as

$$X = \left\langle \left(\sum_{i=1}^{n} \overline{m}_i \langle x_i \overline{m}_i^{-1} \rangle_{m_i} \right) \right\rangle_M \tag{5.5}$$

where $\langle \cdot \rangle$ denotes the modulo operation, X is the binary representation of the number, x_i are its residues, m_i are the moduli, M is the dynamic range, $\overline{m}_i = M/m_i$, and $\overline{m}_i^{-1}$ is the modulo inverse of $\overline{m}_i$. CRT requires the pre-computation of $\overline{m}_i$, and $\overline{m}_i^{-1}$, additions of potentially large products, as well as the final modulo operation with the M, which can be very large. It can be computed, however, in a single cycle. In the other hand, MRC requires the computations of some intermediate coefficients and is a sequential process which requires several steps, but these steps only include small bit-width operations. The Mixed Radix Conversion finds the coefficients $k_1, k_2, \ldots, k_n$, such that

$$X = k_1 + k_2 m_1 + k_3 m_1 m_2 + \cdots + k_n m_1 m_2 \ldots m_{n-1} \tag{5.6}$$

The coefficients are calculated one by one in a number of steps [7], each of which requires the previously calculated coefficients. The modulo inverses can be pre-calculated and pipelining stages can be introduced to make this computation efficient.

5.3.2   Sign Detection

Sign detection is one of the most critical and frequent operation required by NNs, as the Rectified Linear Unit (ReLU), which maps negative values to zero, is the most common activation function. In an RNS representation, numbers in the range $0 < X < R/2$ are positive, whereas numbers in the range $R/2 \leq x < M$ are negative. Magnitude comparison of two RNS numbers which is required for the MaxPooling layers, is also difficult to directly to implement in the RNS domain. Comparison algorithms for particular moduli sets ($2^k - 1, 2^k, 2^k + 1$) [8], or more complex general ones have been proposed [9], that can eliminate the overhead of the conversion. If the choice of moduli is restricted to some specific bases, simple and efficient algorithms have been reported for sign detection [10] and comparison.

5.3.3   Division

Division, which is necessary after the multiplication and accumulation operations of a convolutional layer for example, in order to bring the result in the original dynamic range, also requires special handling. Methods that use special form of moduli, such as powers of two [11] or a product of the moduli [12] as divisors can simplify the hardware implementation. Some methods rely on using small (only one-channel wide) lookup-tables and typically relay on base extension methods. In the case that the divisor is one of the moduli, assuming the number we want to scale is $X \mapsto (x_0, x_1, \ldots, x_{n-1})$, and the modulus to divide by is

602619_1_En_5_Chapter-print ☑ TYPESET ☐ DISK ☐ LE ☑ CP Disp.:26/8/2023 Pages: 94 Layout: German_T5

m_0, we begin by subtracting x_0 from all channels, to obtain the nearest (smaller), number that is divisible by m_0. Because all moduli are co-prime to m_0 there exists a single value k_i for each channel, such that $k_i m_0 \mod m_i = x_i$, which can be obtained using lookup tables. The channel that corresponds to the divisor m_0 requires special handling, since after the subtraction it has a zero value and the value k_0 cannot be directly obtained since $m_0 k_0 \mod m_0$ is always 0, but the residues in channels 2 to n can uniquely define the result. Hence, to obtain the value for the divisor channel from the rest of the channels, a standard *base extension* is needed, which is the process of adding one or more channels to the RNS base. The scaling algorithm is given in Algorithm 1.

Algorithm 1 Scaling by one of the moduli

Require: RNS representation $X = (x_0, x_1, \ldots, x_n)$ and
Require: Base $= (m_0, m_1, \ldots, m_n)$ as inputs.
Require: $K = (k_0, k_1, \ldots, k_n)$ as output.
1: **for** $i \leftarrow 1$ to n **do**
2:    $t_i \leftarrow x_i - x_0 \mod m_i$
3: **end for**
4: **for** $i \leftarrow 0$ to n **do**
5:    $k_i = LUT_{m_0}(t_i)$
6: **end for**
7: $k_0 \leftarrow \text{BE}(k_1, k_2, \ldots, k_n)$ ▷ BE stands for base extension
8: **return** K

5.3.4   Activation Functions

Although the vast majority of the state-of-the-art CNN models only use ReLU as the activation function, other network architectures also make use of other activations (some CNNs do as well). These activations are usually the *hyperbolic tangent* function (tanh) and the *sigmoid* function. For the hardware implementation of these activations several approaches can be used, including lookup table implementations, piecewise linear approximations or low order polynomial approximations [13–15]. Lookup tables are the most straightforward implementation and can be directly adopted in the RNS case as well. However the size of the lookup tables and thus the resulting area overhead can be quite large. Naturally, they result in the minimum approximation error possible. Piecewise linear approximations, on the other hand, are simple and fast but introduce greater error, which, however can be insignificant for many applications [14]. Piecewise linear approximations can also have a simplified RNS implementation.

So, in order to extend the usage of RNS to more complex DNN models such as RNNs, the efficient implementation of other activation functions is required as well. Using the derived

602619_1_En_5_Chapter-print ☑TYPESET ☐DISK ☐LE ☑CP Disp.:26/8/2023 Pages: 94 Layout: German_T5

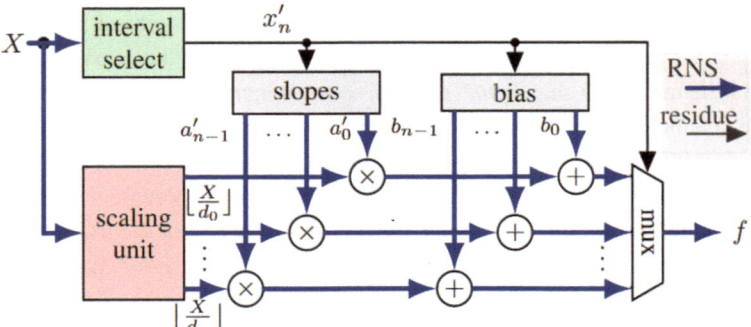

Fig. 5.5 Piecewise linear activation unit

piecewise linear approximations presented in this Section which minimize the maximum approximation error, the common tanh and σ are reduced to scaling and comparison operations. These operations are not straightforward to implement in the RNS domain. A fully-RNS based solution is proposed in [16], requiring no conversions to conventional binary representations. The proposed solution innovates in managing to perform the interval selection required for the piecewise approximations without explicitly performing comparisons which are complex to implement in RNS.

5.3.4.1 Simplified Interval Selection

A general implementation of the piecewise linear activations, shown in Fig. 5.5, requires scaling, a multiplication by some constant and the addition of the bias, to obtain $a_i X + b_i$, where i is the output of the interval selection unit such that $X \in \mathcal{I}_i$, $a_i X$ is obtained as the integer product $a_i' \lfloor \frac{X}{d_i} \rfloor$, $\lfloor \frac{X}{d_i} \rfloor$ is the output of the scaler, and d_i is a product of some of the base moduli. Scaling can be implemented using the algorithms using standard division algorithms [11, 12].

In order to obtain the interval where the input lies, the straightforward approach would be to perform multiple comparisons with the interval edge points. However, since comparison is costly in RNS, an alternative method that segments the dynamic range of the representation into K sub-intervals at once, where K is one of the moduli, can ne used. The method is described by Algorithm 2. Assuming we select the last moduli m_n for this purpose, the algorithm maps the input $X = (x_1, x_2, \ldots, x_n)$ to the *nearest* number (smaller than X) of the form $X' = (0, 0, \ldots, x_n')$ and uses the value of x_n' to decide the interval. To do this a base extension (BE) of the first $n - 1$ channels is performed, to get k which is the last-channel offset between X and X' and is given by $k = (X \bmod \prod_1^{n-1} m_i) \bmod m_n$ Then the value $x_n' = (x_n - k) \bmod m_n$ is obtained and used to distinguish among the various intervals using a look-up table. This process corresponds to calculating $\lfloor \frac{X}{m_1 \cdot m_2 \cdot m_{n-1}} \rfloor$ and separates

602619_1_En_5_Chapter-print ☑TYPESET ☐DISK ☐LE ☑CP Disp.:26/8/2023 Pages: 94 Layout: German_T5

Table 5.1 Approximation parameters

	tanh, $q(x)$			σ, $p(x)$		
i	x_i	a_i	b_i	x_i	a_i	b_i
—	$-\infty$	0	-1	$-\infty$	0	0
0	-1.8148	0.3513	-0.3624	-4.0352	0.0602	0.2428
1	-0.5587	1	0	-1.6529	0.2158	0.5
2	0.5587	0.3513	0.3624	1.6529	0.0602	0.7572
3	1.8148	0	1	4.0352	0	1

Algorithm 2 Interval selection

Require: $X = (x_1, x_2, \ldots, x_n)$
Ensure: the index i, such that $X \in \mathcal{I}_i$
1: $k = \mathrm{BE}(m_n, (x_1, x_2 \ldots, x_{n-1}))$ ▷ base extension
2: $x_n' = (x_n - k) \mod m_n$ ▷ computes x_n'
3: $i = \mathrm{LUT}(x_n')$ ▷ obtain lookup table to generate index i

---

the interval into $K = m_n$ sub-intervals, defined by the integer multiples of $\prod_{i=1}^{n-1} m_i$. The algorithm could be used for piecewise approximations with up to 32 intervals (Table 5.1).

5.3.4.2 Sigmoid and tanh Approximations

Approximations of the common activations of tanh and sigmoid which combine hardware simplicity and sufficient accuracy for the application are required for an efficient RNS implementation. Among the wide variety of techniques to approximate a function $f(x)$ [17], one can minimize the maximum absolute error $|\varepsilon|$ by partitioning the domain of $f(x)$ into successive intervals, $\mathcal{I}_i$, by a sequence of points $\{x_i : i = 0, 1, \ldots, N-1\}$, i.e., $\mathcal{I}_i = [x_i, x_{i+1}]$, and approximating $f(x)$, $x \in \mathcal{I}_i$, by $p_i(x) = a_i x + b_i$. Error minimization is achieved by requiring the same $|\varepsilon|$ for all $\mathcal{I}_i$, with alternating error signs in successive intervals and at the interval boundaries. Let the maximum error in $\mathcal{I}_i$ be observed at point z_i, $x_i < z_i < x_{i+1}$, then we require

$$f'(z_i) - a_i = 0 \tag{5.7}$$

$$f(x_i) - a_i x_i - b_i = (-1)^i |\varepsilon| \tag{5.8}$$

$$f(z_i) - a_i z_i - b_i = (-1)^{i+1} |\varepsilon| \tag{5.9}$$

Additional continuity constraints are imposed at $\mathcal{I}_i$ boundaries, i.e., $p_i(x_i) = p_{i+1}(x_i)$. By numerically solving the nonlinear system of constraints for $a_i, b_i, x_i, z_i, \varepsilon$, for all i, the approximations shown in Fig. 5.6 as dark blue lines are derived, achieving the errors $e_p(x)$.

602619_1_En_5_Chapter-print ☑TYPESET ☐DISK ☐LE ☑CP Disp.:26/8/2023 Pages: 94 Layout: German_T5

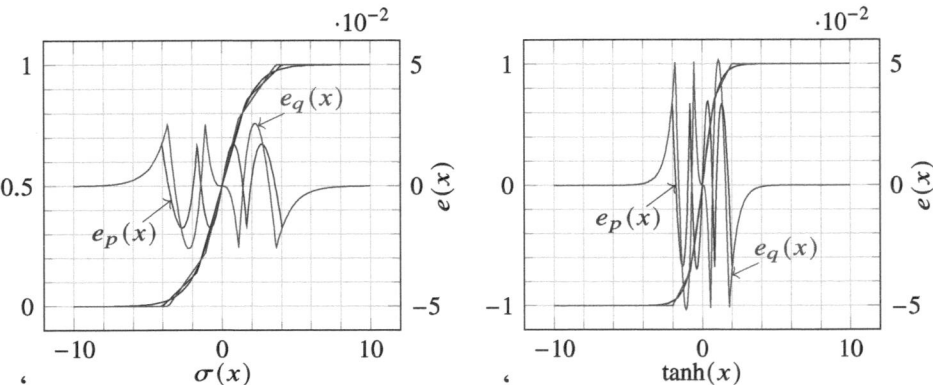

Fig. 5.6 Piecewise-linear activation approximations and errors for five intervals

By constraining the slopes and biases in the central interval $[x_1, x_2]$ to 1 and 0 for tanh and $\frac{1}{4}$ and $\frac{1}{2}$ for σ, and solving for the remainder of parameters, the error in the vicinity of $x = 0$ is small; however the maximum error is worse, as shown by $e_q(x)$ in Fig. 5.6.

5.4   Partially RNS-Based Architectures

A common approach in RNS-based DNN implementations is to perform all multiply-add operations of a single convolutional or dense layer in the RNS representation and then use a converter to obtain a partial result in normal positional binary representation [18–21]. With this intermediate result, the non-linear activation functions (*ReLU*, tanh, *softmax*) can be computed and the results can be again converted to RNS format to be fed to the next layer.

Many application-specific AI accelerator designs, as well as more general-purpose architectures, such as TPUs or GPUs, perform DNN computations by decomposing them into the matrix or vector multiplication primitives. Thus, by utilizing efficient hardware matrix multipliers, performance can be orders of magnitude better than CPUs. An RNS TPU (Tensor Processing Unit) is proposed in [18]. In the core of this architecture, there is an RNS matrix multiplier implemented as a two-dimensional systolic array. Each processing element performs one operation (MAC) at each cycle and passes the result to neighboring processing elements. Systolic arrays are an efficient way of increasing throughput and dealing with the limited memory bandwidth problem. In this particular RNS systolic array, each processing element decomposes the larger MAC operation (typically 8 or 16 bits), into smaller, each within the range of the respective channel, that can be performed in parallel. Using an FPGA implementation the RNS matrix multiplier is reported to perform a 32×32 fixed point matrix multiplication up to $9\times$ more efficiently than a binary matrix multiplier for large matrices.

602619_1_En_5_Chapter-print ☑ TYPESET ☐ DISK ☐ LE ☑ CP Disp.:26/8/2023 Pages: 94 Layout: German_T5

In [20] the authors extend the RNS usage to the implementation of the convolution operation. Individual layers are executed on an RNS-based FPGA accelerator. However results are sent to a CPU, which performs the non-trivial RNS operations, such as applying the activation functions, before being sent back to the FPGA for the execution of the next layer. RNS results in a 7.86–37.78% reduction of the hardware costs of a single convolutional layer compared to the two's-complement implementation, depending on the RNS base selection.

A variant of the Residue Number System, called the *Nested RNS* (NRNS) is proposed in [19]. NRNS applies a recursive decomposition of the residue channels into smaller ones. Adder and multipliers can be thus implemented by using smaller and faster circuits. Assuming that a number X has a RNS representation of $(x_1, x_2, \ldots, x_n)$, then the nested RNS representation will be of the form

$$X = (x_1, x_2, \ldots, (x_{i1}, x_{i2}, \ldots, x_{im}), \ldots, x_n) \tag{5.10}$$

where $(x_{i1}, x_{i2}, \ldots, x_{im})$ is the RNS decomposition of the $i - th$ channel This technique introduces an additional complexity, as any operation must be recursively applied to each level of the representation, however it manages to handle large dynamic ranges with very small channels. The authors use a 48-bit equivalent dynamic range composed only of 4-bit MAC units which can be realized by look-up tables of the FPGA. Contrary to [20], which relies on an external CPU, in this work binary-to NRNS and NRNS-to-binary conversions are realized by DSP blocks and on-chip BRAMs. After Input data are converted into the NRNS representation, a number of parallel convolutional units perform all the necessary computations of a single convolutional layer. The results are then converted to binary using a tree-based NRNS-to-binary converter. The authors report a performance per area improvement of $5.86\times$ compared to state-of-the-art FPGA implementations for the ImageNet benchmark. In a different approach the RNS arithmetic costs are reduced by restricting the RNS base selection to low-cost moduli of the form $2^k \pm 1$ [21]. This way, modified fast prefix adders and CSA trees using end-around-carry propagation can be used, diminishing any overhead of the modulo operator.

In another category of RNS-based architectures, the usage of very small channels allows the realization of multiplier-free CNN architectures. The authors utilize a small RNS base of (3,4,5) and reduce the implementation of the multiplications to shifts and additions [22]. Despite the reduced dynamic range of the representation, the authors report minimal accuracy loss, while achieving 36 and 23% reduction in power and area, respectively. A method to drastically reduce the number of multiplications in CNN RNS-bases accelerators is proposed in [23]. It utilizes a modified hardware mapping of the convolution algorithm where the order of operations is rearranged. Because of the small dynamic range of each RNS channel, there is an increased number of common factors inside the weight kernels during convolution. By first executing the additions of the input feature map terms that correspond to the same factors, and then performing the multiplications with the common weight factors, a 97%, reduction of the total multiplications is reported for state-of-the-art CNN models.

602619_1_En_5_Chapter-print ☑TYPESET ☐DISK ☐LE ☑CP Disp.:26/8/2023 Pages: 94 Layout: German_T5

5.5   End-to-End RNS Architectures

While the above circuits mange to achieve some performance gain in the implementation of a single convolutional layer, they require significant amounts of extra hardware to perform the conversions which can become the bottleneck for some of these designs. More recent approaches focus on overcoming the difficulties of performing operations such as sign detection, comparison, and scaling which is usually required following multiplication. In these approaches, input data are initially converted to an RNS representation and then the entire processing takes place in the RNS domain.

5.6   State-of-the-Art End-to-End RNS Architectures

The system in [24] introduces some novel mechanisms for dealing with this problem and proposes an efficient fully RNS-based architecture. The authors of this work choose to work with moduli of the form $2^k - 1$, 2^k, $2^{k+1} - 1$. In particular they select (31,32,63) as the basis of their representation, as it is found to provide a sufficient dynamic range (16-bit equivalent), that results in no accuracy loss, for state-of-the-art networks and benchmarks. For the design of the modulo adders, which are simplified due to the particular selection of the moduli, parallel-prefix Sklansky adders with an end around carry are utilized. For the multiplications, a radix-4 Booth encoding is adopted within each channel. An optimized sign detection unit for this set of moduli is used, based on an approach proposed in [10]. which can be further transformed and result in a relatively hardware-friendly implementation. Using a similar logic to the work proposed in [8], the comparison of two RNS numbers can also be implemented by calculating auxiliary partitioning functions.

The authors also introduce a base extension mechanism which is necessary in order to avoid potential overflow when accumulating the partial sums. In this work, a base extension method proposed in [25] is used, where the middle channel is extended from 2^k to 2^{k+e}. This way the convenient properties of the chosen moduli are maintained. Base extension takes places once before each multiplication to ensure that the product lies within the dynamic range and then again before the accumulation. The authors define the number of extra bits that are added each time based on extensive simulation on benchmark networks and on a per-layer basis. RNS circuits result in significant delay and energy efficiency improvement, especially in the case of multiplication at the cost of larger overall area. Comparisons in terms of various performance metrics against the Eyeriss [26] accelerator are reported for various networks. Up to 61% reduction in energy consumption compared to the conventional positional binary representation has been achieved. The system can also support an increased clock frequency as high as 1.20 GHz versus 667 MHz in the case of the positional binary system, indicating a 1.8× improvement in computational latency.

5.7   In-Memory Computing RNS Architectures

Recently, there has been a growing focus of AI accelerator design research on in-memory computing. This is because of the paradigm-shifting effect that emerging memory technologies can have on processing systems. It is known that the largest part of the energy consumption of any DNN accelerator is due to the memory accesses and data transferring, particularly to and from the off-chip RAM. In-memory computing (IMC) aims to diminish data transfer costs by bringing the computing inside or near the memory elements.

Efforts have been made to bring the benefits of the RNS to IMC systems. In these (mainly digital) IMC designs, the benefit of the usage of RNS over binary representation stems from the speedup of the bitwise serial addition operations, due to the inherent parallel operations of the RNS channels.

RNS has been utilized in the design of an in-memory computing system [27]. In this work, the selected moduli are of the form $2^k - 1, 2^k, 2^k + 1$. A sign detection mechanism similar to [10], is developed, in order to implement the ReLU and MaxPooling operations without having to convert to a binary representation. Addition and multiplication within each RNS channel, take place inside the memory elements. Multiplication of two numbers, a, b is implemented through addition and memory accesses by calculating the quantity $\frac{(a+b)^2}{4} - \frac{(a-b)^2}{4}$, where squaring is implemented using look-up-tables. A single crossbar memory is assigned to each neuron and supports in-memory addition in a tree-based structure. For this purpose, a Memristor Aided loGIC (MAGIC) is used. Based on experimental results, the proposed RNS in-memory architecture consumes $145.5\times$ less energy and leads to a speedup of $35.4\times$ compared to NVIDIA GPU GTX 1080.

An near-memory RNS-based processing architecture is proposed in [28]. Instead of memristor-based memory macros, a DRAM computational sub-array is utilized for the implementation of the MAC operations in the RNS domain, combined with parallel-prefix adders, to implement bitwise multiplication and accumulation. Unlike [24], where multiplication is directly implemented in memory (by mapping to additions and squaring), here they are implemented by combining elementary bit-wise operations (AND, OR, XOR) between the operands. The authors also design a more flexible activation function unit which is based on a Mixed-Radix conversion. Similar to [24], an RNS base of $(2^k - 1, 2^k, 2^{k+1} - 1)$ is utilized. Gains in the order of $331 - 897\times$ in terms of energy efficiency compared to GPU platforms are reported, and $2\times$ compared to other IMC designs.

5.8   Summary of RNS-Based DNN Architectures

RNS-based architectures targeting DNN applications are summarized in Table 5.2. The majority of these approaches utilize low-cost moduli of the form $2^k - 1, 2^k, 2^k + 1$ to reduce the overhead of the modulo operator and are targeting CNNs. Most of these RNS accelerators can achieve speedups in the order of $1.5 - 3\times$ and can also be more energy

Table 5.2 RNS-based architectures targeting DNN applications.

References	Application	RNS Arithmetic Details	RNS Base	Dataset	NN Model	Platform	Accuracy drop (%)	Speedup (×)	Energy reduction (%)
[19]	CNN	Partialy—Nested	No specific	ImageNet	ConvNet2	FPGA	–	3.1	–
[22]	CNN	Partialy—Mult. free	$\{3,4,5\}$	MNIST, CIFAR10, ImageNet	LeNet, VGG, AlexNet, Resnet50	FPGA	0.03	2.8	36
[24]	CNN	End-to-end	$\{31,32,63\}$	ImageNet	AlexNet, VGG-16, SqueezeNet}	ASIC	0.27	2.9	30–61
[27]	CNN	End-to-end (in-memory)	$\{2^k - 1, 2^k, 2^k + 1\}$	MNIST, INDOOR, CIFAR10	Custom	Memristor IMC	1	1.6–5	500–840
[28]	CNN	End-to-end (in-memory)	$\{31,32,63\}$	MNIST,SVHN, CIFAR10	DoReFa-Net	DRAM IMC	1	2–4.3	90–100
[16]	LSTM	End-to-end	$\{3,5,7,31,32\}$	Q-traffic	Custom	ASIC	3	1.6	30

602619_1_En_5_Chapter-print ☑ TYPESET ☐ DISK ☐ LE ☑ CP Disp :26/8/2023 Pages: 94 Layout: German_T5

efficient. IMC RNS-based systems exhibit the largest energy savings. Among conventional systems, [24] illustrates more clearly the applicability of the RNS in DNN architectures by proposing a fully RNS system that outperforms the binary state-of-the-art counterpart. The RNS usage is also extended to LSTM networks, by designing hardware-friendly RNS activation units for the implementation of $tanh$ and $sigmoid$ functions [16].

In conclusion, the Residue Number System can be an attractive number representation choice for DNN accelerators, and several RNS-based architectures have been reported recently targeting AI applications, due to its various advantages. RNS exhibits inherent parallelism at the residue channel processing level. It utilizes parallel computations along separate residue channels, where operations in each of them are performed modulo a specific modulus, with no need for information (carry or other) to be shared between residue channels.

The main challenge in designing an efficient RNS-based accelerator is to minimize or, possibly, eliminate the overhead introduced due to the implementation of non-linear operations. Another key factor is the optimization of the moduli selection and the corresponding arithmetic circuits, to meet the accuracy requirements.

Some of the RNS systems proposed in recent literature only perform the MAC or matrix multiplication operation, required by the convolutional layers, in the RNS representation, and use intermediate converters between number systems for the non-linear operations. More recently, completely RNS-based approaches have been proposed that eliminate the overhead introduced by these intermediate conversions to and from a traditional positional binary representation.

References

1. Soderstrand, M.A., Jenkins, W.K., Jullien, G.A., Taylor, F.J. (eds.): Residue Number System Arithmetic: Modern Applications in Digital Signal Processing. IEEE Press (1986)
2. Nannarelli, A., Re, M.: Residue Number Systems: a Survey. No. 2008-04 in D T U Compute. Technical Report. Technical University of Denmark, DTU Informatics, Building 321 (2008)
3. Vergos, H., Efstathiou, C., Nikolos, D.: Diminished-one modulo $2^n + 1$ adder design. IEEE Trans. Comput.—TC **51**, 1389–1399 (2002)
4. Paliouras, V., Karagianni, K., Stouraitis, T.: A low-complexity combinatorial RNS multiplier. IEEE Trans. Circuits Syst. II: Analog Digit. Signal Proc. **48**, 675–683 (2001). https://doi.org/10.1109/82.958337
5. Kong, Y., Asif, S., Khan, M.: Modular multiplication using the core function in the residue number system. In: Applicable Algebra in Engineering, Communication and Computing, pp. 1–16 (2016)
6. Efstathiou, C., Vergos, H., Dimitrakopoulos, G., Nikolos, D.: Efficient diminished-1 modulo $2^n + 1$ multipliers. IEEE Trans. Comput.—TC **54**, 491–496 (2005)
7. Gbolagade, K., Cotofana, S.: An O(n) Residue Number System to Mixed Radix Conversion Technique, pp. 521–524 (2009). https://doi.org/10.1109/ISCAS.2009.5117800
8. Torabi, Z., Jaberipur, G.: Low-power/cost RNS comparison via partitioning the dynamic range. IEEE Trans. Very Large Scale Integr. (VLSI) Syst. **24**(5), 1849–1857 (2016). https://doi.org/10.1109/TVLSI.2015.2484618

9. Xiao, H., Ye, Y., Xiao, G., Kang, Q.: Algorithms for comparison in residue number systems. In: Asia-Pacific Signal and Information Processing Association Annual Summit and Conference (APSIPA), pp. 1–6 (2016). https://doi.org/10.1109/APSIPA.2016.7820790

10. Xu, M., Bian, Z., Yao, R.: Fast sign detection algorithm for the RNS moduli set $\{2^{n+1} - 1, 2^n - 1, 2^n\}$. IEEE Trans. Very Large Scale Integr. (VLSI) Syst. **23**(2), 379–383 (2015). https://doi.org/10.1109/TVLSI.2014.2308014

11. Meyer-Base, A., Stouraitis, T.: New power-of-2 RNS scaling scheme for cell-based IC design. IEEE Trans. VLSI Syst. **11**, 280–283 (2003). https://doi.org/10.1109/TVLSI.2003.810799

12. Kong, Y., Phillips, B.: Fast scaling in the residue number system. IEEE Trans. VLSI Syst. **17**, 443–447 (2009). https://doi.org/10.1109/TVLSI.2008.2004550

13. Namin, A., Leboeuf, K., Wu, H., Ahmadi, M.: Artificial Neural Networks Activation Function HDL Coder, pp. 389–392 (2009). https://doi.org/10.1109/EIT.2009.5189648

14. Nicolae, A.: Plu: The Piecewise Linear Unit Activation Function (2008). arXiv:1809.09534

15. Kundu, A., Heinecke, A., Kalamkar, D., Srinivasan, S., Qin, E.C., Mellempudi, N.K., Das, D., Banerjee, K., Kaul, B., Dubey, P.: K-tanh: Efficient tanh for Deep Learning (2019). arXiv:1909.07729

16. Sakellariou, V., Paliouras, V., Kouretas, I., Saleh, H., Stouraitis, T.: A High-performance RNS LSTM block. In: IEEE International Symposium on Circuits and Systems (ISCAS) (2022)

17. Muller, J.M.: Elementary Functions: Algorithms and Implementation. Birkhäuser (1997)

18. Olsen, E.: RNS Hardware Matrix Multiplier for High Precision Neural Network Acceleration: RNS TPU, pp. 1–5 (2018). https://doi.org/10.1109/ISCAS.2018.8351352

19. Nakahara, H., Sasao, T.: A deep convolutional neural network based on nested residue number system. In: 2015 25th International Conference on Field Programmable Logic and Applications (FPL), pp. 1–6 (2015). https://doi.org/10.1109/FPL.2015.7293933

20. Valueva, M., Nagornov, N., Lyakhov, P., Valuev, G., Chervyakov, N.: Application of the residue number system to reduce hardware costs of the convolutional neural network implementation. Math. Comput. Simul. **177**, 232–243 (2020)

21. Abdelhamid, M., Koppula, S.: Applying the Residue Number System to Network Inference (2017). arXiv:1712.04614

22. Salamat, S., Shubhi, S., Khaleghi, B., Rosing, T.: Residue-Net: Multiplication-free neural network by in-situ no-loss migration to residue number systems. In: 2021 26th Asia and South Pacific Design Automation Conference (ASP-DAC), pp. 222–228 (2021)

23. Sakellariou, V., Paliouras, V., Kouretas, I., Saleh, H., Stouraitis, T.: On reducing the number of multiplications in RNS-based CNN accelerators. In: IEEE International Conference on Electronics, Circuits, and Systems (ICECS), pp. 1–6 (2021). https://doi.org/10.1109/ICECS53924.2021.9665461

24. Samimi, N., Kamal, M., Afzali-Kusha, A., Pedram, M.: Res-DNN: a residue number system-based DNN accelerator unit. IEEE Trans. Circuits and Syst. I: Regul. Pap. **67**(2), 658–671 (2020). https://doi.org/10.1109/TCSI.2019.2951083

25. Szabó, N.S., Tanaka, R.I.: Residue Arithmetic and its Applications to Computer Technology (1967)

26. Chen, Y.H., Emer, J., Sze, V.: Eyeriss: A Spatial Architecture for Energy-Efficient Dataflow for Convolutional Neural Networks (2016). https://doi.org/10.1145/3007787.3001177

27. Salamat, S., Imani, M., Gupta, S., Rosing, T.: RNSnet: In-memory Neural Network Acceleration using Residue Number System, pp. 1–12 (2018)

28. Roohi, A., Taheri, M., Angizi, S., Fan, D.: RNSiM: efficient deep neural network accelerator using residue number systems. In: IEEE/ACM International Conference On Computer Aided Design (ICCAD), pp. 1–9 (2021). https://doi.org/10.1109/ICCAD51958.2021.9643531

602619_1_En_5_Chapter-print ☑ TYPESET ☐ DISK ☐ LE ☑ CP Disp.:26/8/2023 Pages: 94 Layout: German_T5

BFP for DNN Architectures

6

Abstract

This chapter explores the use of the block floating point number system as an alternative to traditional fixed point and floating point number systems in deep neural network implementations. We begin by providing an overview of BFP and explaining how it differs from FLP and FXP. Next, we discuss the factors that can impact the performance of BFP in DNN acceleration, drawing from existing literature to identify the best practices for optimizing BFP performance. Finally, we present quantitative comparisons of various works that have utilized BFP for DNNs, demonstrating its potential as an effective tool for accelerating DNN computations.

6.1 BFP Overview

BFP representation offers a middle ground between FLP and FXP formats. This representation is proposed to preserve accuracy comparable to full precision FLP and hardware efficiency comparable to FXP. This is achieved by representing numbers with an exponent and a mantissa similar to FLP to guarantee a wide dynamic range. However, instead of representing each value separately, a group (called here a block) of values has a common exponent while maintaining private mantissas, as we can see in Fig. 6.1. Let N be a tensor that represents a block of t elements initially represented in FLP as

$$N = (n_1, \ldots n_i, \ldots n_t),$$

$$= ((-1)^{s_1} m_1 2^{e_1}, \ldots (-1)^{s_i} m_i 2^{e_i}, \ldots (-1)^{s_t} m_t 2^{e_t}). \tag{6.1}$$

© The Author(s), under exclusive license to Springer Nature Switzerland AG 2024
G. Alsuhli et al., *Number Systems for Deep Neural Network Architectures*,
Synthesis Lectures on Engineering, Science, and Technology,
https://doi.org/10.1007/978-3-031-38133-1_6

61

602619_1_En_6_Chapter-print ☑ TYPESET ☐ DISK ☐ LE ☑ CP Disp.:26/8/2023 Pages: 94 Layout: German_T5

Fig. 6.1 The bit representation
for the block floating point
number system

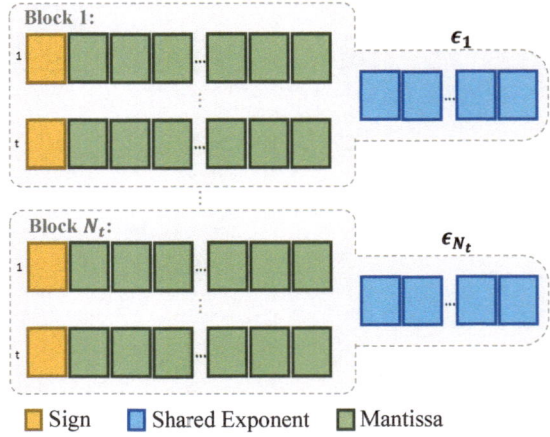

This block is represented with BFP format as $\grave{N}$ such that

$$\grave{N} = (\grave{n}_1, \ldots \grave{n}_i, \ldots \grave{n}_t), \tag{6.2}$$
$$= ((-1)^{s_1} \grave{m}_1, \ldots (-1)^{s_i} \grave{m}_i, \ldots (-1)^{s_t} \grave{m}_t) \times 2^{\epsilon_N},$$

where ϵ_N is a shared exponent between the elements of block N, and $\grave{m}_i$ is the aligned mantissa of element i such that $\grave{m}_i = \mathbf{BS}(m_i, e_i - \epsilon_N)$, where $\mathbf{BS}$ is the bit-shift operation. For the large difference between the Private and shared exponents ($e_i - \epsilon_N$), this shifting causes some of the least-significant bits of the mantissa to be truncated. The truncation happens frequently when there are many outliers in a block, which in turn depends on the size of the block and the way the shared exponent is selected.

6.2   BFP for DNNs

Since the dot product is the basic operation involved in DNN inference and training, the main target of BFP is to simplify the complex hardware required to perform this operation when FLP is used. For two blocks $\grave{N}_1$ and $\grave{N}_2$ represented in BFP, the dot product is calculated as

$$\grave{N}_1.\grave{N}_2^T = ((-1)^{s_{1,1}} \grave{m}_{1,1}, \ldots (-1)^{s_{t,1}} \grave{m}_{t,1}) \times 2^{\epsilon_{N_1}}. \tag{6.3}$$
$$((-1)^{s_{1,2}} \grave{m}_{1,2}, \ldots (-1)^{s_{t,2}} \grave{m}_{t,2})^T \times 2^{\epsilon_{N_2}}$$
$$= 2^{\epsilon_{N_1} + \epsilon_{N_2}} \sum_{i=1}^{t} (-1)^{s_{i,12}} \grave{m}_{i,1} \times \grave{m}_{i,2},$$

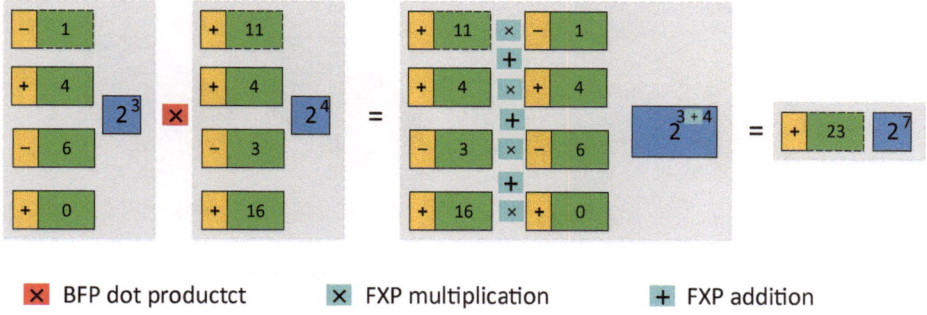

Fig. 6.2 The decomposition of the dot product between two BFP groups into FXP multiplication and addition

where $s_{i,j}$ and $m_{i,j}$ are the sign and mantissa of the ith element in the jth block, respectively, ϵ_{N_j} is the shared exponent of the jth block, $s_{i,12}$ results from XORing $s_{i,1}$ and $s_{i,2}$, and T stands for transportation. Equation (6.3) shows that the dot product of two blocks of size t represented in BFP involves t FXP multiplications of mantissas, $t-1$ FXP additions of the products, and one addition of the two shared exponents. Figure 6.2 illustrates the dot product between two groups represented in BFP. The additional overhead compared to FXP representation comes from the hardware required to handle the shared exponent which mainly depends on the number of the blocks [1]. As a result, the performance of DNN in the presence of BFP representation is determined by block partition scheme, shared exponent selection, and the bit-width of the mantissa and shared exponent, which will be discussed next.

6.2.1 BFP Block Design

Determining how the blocks are partitioned is essential to achieving good DNN performance with BFP [2, 3]. Usually, the input activation of each layer is considered as one block, whereas the weight matrix needs a specific scheme to be divided into blocks. There are two known blocking approaches, filter-based blocking [2–8] and tile-based blocking [1, 9–13], illustrated in Fig. 6.3a and b, respectively. In the filter-based blocking, each filter of weights along the input channels is considered a block. Then the total number of blocks equals to the number of filters. This blocking is usually called coarse-grain blocking and it is the most hardware-friendly blocking approach as the accumulation of each output activation is done with the same shared exponent. Thus, it can be done using the FXP arithmetic [2]. However, this approach may end up with severe accuracy degradation due to the increased number of outliers that need to be truncated within these large blocks.

On the other hand, tile-based blocking is proposed to strike a compromise between accuracy and hardware efficiency. This approach relies on breaking large matrices of the

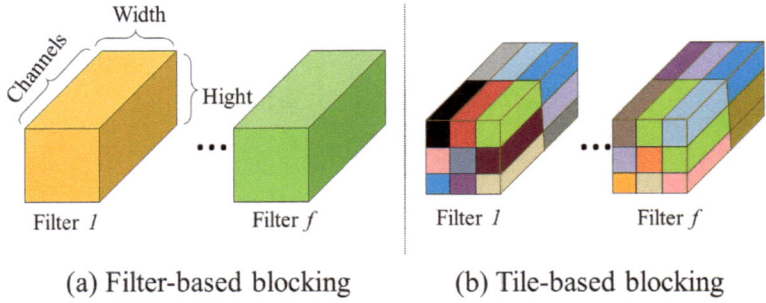

(a) Filter-based blocking   (b) Tile-based blocking

Fig. 6.3 Different blocking schemes, different colors indicate differed blocks

filters down into small tiles to fit into limited hardware resources. Each tile is considered as a block with a shared exponent. The size of these tiles is a metric that need to be optimized. For example, a large tile of size 576 is used in [12] which requires a 12-bit mantissa to obtain acceptable accuracy. However, the authors in [11] showed that 12-bit FXP can achieve similar accuracy with simpler hardware implementation. This indicates that BFP may has no advantage over FXP for such large tiles. Smaller tiles of 16 elements are used in [1, 9] seeking better accuracy, but with an added hardware complication comes from the need to convert to FLP before the accumulation.

6.2.2 Shared Exponent Selection

One shared exponent for each block needs to be selected after partitioning the blocks[1] and whenever a new block is created with multiple shared exponents. For example, this exponent is aligned after performing the calculation of each DNN layer as the calculation of the output activation usually ends up with a matrix of multiple exponents [3]. To this end, most of DNN accelerators that adopt BFP calculate the shared exponent dynamically during DNN training or inference. Static shared exponent selection can be utilized prior to DNN inference.

One of two schemes is usually used for this dynamic shared exponent selection; maximal exponent-based or statistics-based schemes. The dynamic maximal exponent selection scheme is more popular [14–16]. In this scheme, for each block of (6.1), different floating point numbers n_i are compared and the maximum exponent is selected as follows

$$\epsilon_N = \max_{e^i} : i \in 1, \ldots, t. \tag{6.4}$$

To find this maximal exponent before performing the dot product between weights and activations resulting from the previous layer, the output activations represented in BFP with

---

[1] As the case of partitioning the weight blocks prior to DNN inference, which is usually performed offline.

602619_1_En_6_Chapter-print ☑ TYPESET ☐ DISK ☐ LE ☑ CP Disp.:26/8/2023 Pages: 94 Layout: German_T5

several exponents need to be converted back into FLP, which adds a large overhead on the performance and the resources.

To keep the advantage of the dynamic calculation of the shared exponent while avoiding frequent conversion between BFP and FLP the statistics-based scheme is proposed to predict the shared exponent during DNN training [8, 17]. In this scheme, the optimal exponent for each block is predicted based on statistics collected in the previous learning iteration. For example, in [8] the maximum value recorded within each block is stored for the last i iterations. Then, the maximum and the standard deviation of the stored values are used to calculate the shared exponent for the next iteration. This scheme works because the values within each block change slowly during the training. However, although this scheme avoids the conversion to FLP to calculate the exponent, some additional overhead is required to store the recorded statistics for each block. Thus, this scheme is suitable for the case when the number of blocks is relatively small.

The static shared exponent scheme is presented to get rid of exponent calculation overhead when the BFP is employed for CNN inferences rather than training [1, 5, 6, 18]. Instead of dynamically calculating the shared exponent during run-time, the shared exponent can be set to a constant value estimated offline. The common approach to determine the shared exponent offline is to minimize the Kullback-Leibler (K-L) divergence [19] between FLP32 distribution and BFP distribution of all blocks before the inference. By doing so, the extra memory and computational resources used for the exponent and the conversion between BFP and FLP are eliminated [1]. Because the input and output activations may have different shared exponents, a bit-shifting is needed after each layer calculation, Fig. 6.4c. Figure 6.4 summarizes the dataflow of the BFP when each of the three shared exponent determination schemes is adopted.

6.2.3   BFP Precision

The precision of BPF is determined by the number of bits allocated for both the shared exponent and mantissa. Reducing this precision is an objective to increase arithmetic efficiency and decrease the required memory. At the same time, the over-reduced bit-width of the mantissa results in what is known as a zero setting problem [20]. This problem occurs when all the bits of the mantissa are shifted out resulting in a zero number representation, despite the presence of the exponent value.[2] The over-reduction of the shared exponent number of bits is much worse. This is because of insufficiency to represent the actual exponent of the block, and thus the caused truncation ruins the correct representation of all numbers in the block. BFP precision is usually either static [1–8, 10, 12–14, 17, 18, 21, 22] or dynamic [11, 20].

---

[2] One solution for the zero setting problem is using stochastic rounding i.e. adding stochastic noise after shifting the mantissas. The added noise help in solving this problem and allowed for further reduction of the mantissa bits while preserving high accuracy in [11].

602619_1_En_6_Chapter-print ☑ TYPESET ☐ DISK ☐ LE ☑ CP Disp.:26/8/2023 Pages: 94 Layout: German_T5

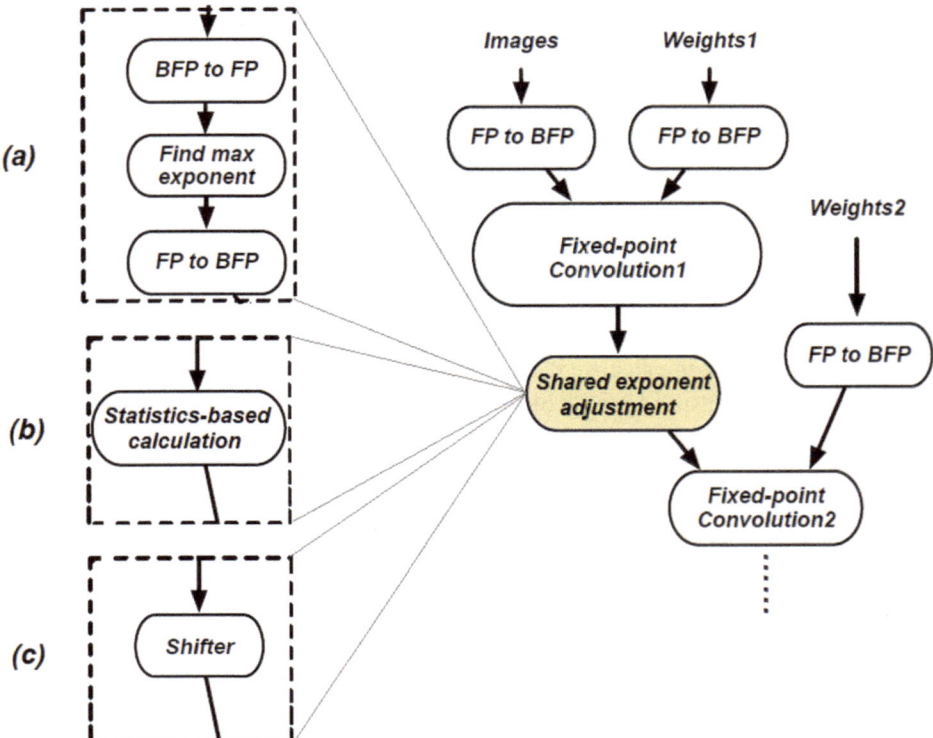

Fig. 6.4 Shared exponent adjustment schemes: **a** Dynamic based on maximal exponent **b** Dynamic based on statistics **c** Static

6.2.3.1 Static Precision

In static precision, the number of bits is fixed and selected offline. To select the best precision, usually, a few experiments are performed using a different number of bits [4, 5, 17]. This gives an insight into the impact of this metric on the performance of DNN and allows for picking the minimum number of bits that preserve acceptable accuracy or the one that gives the best trade of between hardware efficiency and accuracy. Reducing the mantissa bit-width was paying attention in the literature because the performance of DNN is less sensitive to mantissa reduction compared to the shared exponent. For example, 23-bit mantissa, same as the case of FLP, is required to guarantee the convergence of the Q-learning in [17], whereas 8-bit mantissa, or even less, was found to be sufficient for other CNN accelerator designs [1, 1–7, 11, 13, 18, 20–22]. This indicates that the required static precision depends on the problem to be solved (mainly, the used dataset and DNN model).

602619_1_En_6_Chapter-print ☑TYPESET ☐DISK ☐LE ☑CP Disp.:26/8/2023 Pages: 94 Layout: German_T5

6.2.3.2 Dynamic Precision

Dynamic precision means that the number of bits allocated to the represented values is not constant and can change over time. Usually, the mantissas rather than the exponents are assigned dynamic precisions [11, 20]. This dynamic precision is basically needed when the implemented DNN architecture is intended to be used for training rather than inference. This is attributed to the fact that the distributions of the weights, activations, and weight updates change during the training. Figure 6.5 [23] is an example of how the distributions of

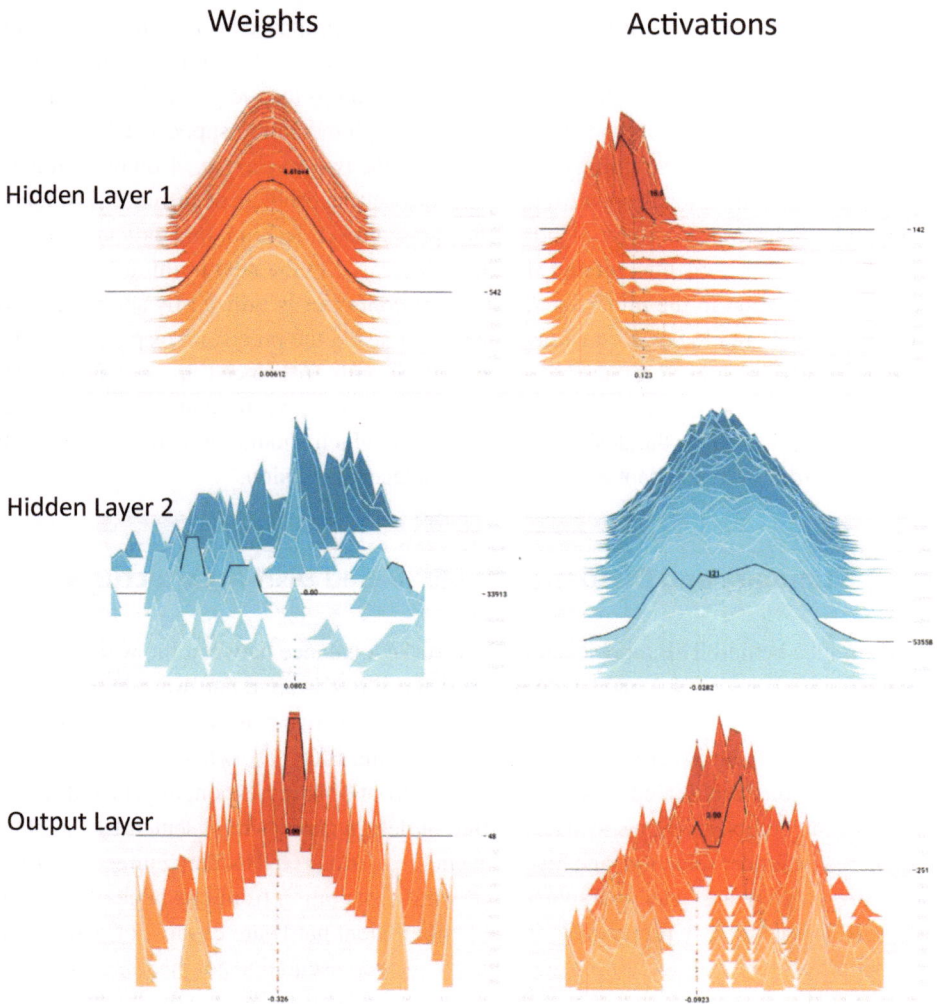

Fig. 6.5 Example of how the distribution of weights and activations changes during different DNN training steps. These histograms illustrate how the values of weights and activation change for particular layers over time. The x-axis shows the actual value and the y-axis shows training steps [23]

602619_1_En_6_Chapter-print ☑ TYPESET ☐ DISK ☐ LE ☑ CP Disp.:26/8/2023 Pages: 94 Layout: German_T5

the weights and activations change during DNN training and from layer to layer. To speed up the training, the authors in [11] proposed adaptive training by changing the precision of BFP progressively across both training iterations and layer depth. This relies on the fact that the training is more amenable to low precision in its early stages. In their approach, two levels of precision are supported, mainly 4-bit and 2-bit mantissas. For each block, the relative improvement due to using the higher precision is estimated by quantizing the block numbers using both precisions. Then, if this relative improvement is higher than a threshold the higher precision is used. This threshold differs based on the layer depth and training iteration.

On the other hand, mixed dynamic precision of BFP is proposed because the distribution of the weight updates (gradients) changes more frequently than other variables during training [20]. This scheme assigns different, higher, precision to the weight updates compared to weights and activations. At the same time, their implementation supports adjusting this precision online during the training time to be one of the two levels (e.g., 4-bit or 8-bit mantissa). For each training iteration, the number of zero-setting problem occurrences is tracked. If this problem happens more frequently than a predefined threshold, this indicates that the current precision is not sufficient and should be increased in the next training iteration. To avoid the fluctuation in the precision, a hysteresis controller is utilized by specifying two thresholds, upper and lower, for increasing and decreasing the precision, respectively. This dynamic precision showed no accuracy degradation with 16% speed-up compared to the static precision. However, the dynamic precision advantage usually comes at the expense of added complication to the design of the hardware which should be reconfigurable with multi-mode arithmetic to adapt according to the selected precision.

6.3   Summary and Discussion of BFP-Based DNN Architectures

The main idea behind BFP representation is to strike a balance between the wide dynamic range but hardware-inefficient FLP format and the limited-range hardware-friendly FXP format. BFP can be considered as a general format that has two extreme cases, i.e., the FLP case when each value is set in a separate block and the FXP case when the whole values of the architecture are treated as a single block with one shared exponent. Thus, different trade-offs can be obtained by specifying different design choices represented by the block size, shared exponent selection, and bit-width choice. Various CNN architectures that utilize BFP representation are listed in Table 6.1. The first observation from this table is that even though BFP was initially proposed to implement efficient hardware capable of performing CNN training phase without ruining the accuracy, this representation got the same amount of attention for highly accurate inference hardware implementation. Most of these architectures achieved negligible accuracy degradation compared to FLP even with less than 8-bit mantissa [1, 11, 22]. Different implementations make use of different combinations of the discussed design choices, thus, the reported results of these works can't be used to prove the superiority

602619_1_En_6_Chapter-print ☑ TYPESET ☐ DISK ☐ LE ☑ CP Disp.:26/8/2023 Pages: 94 Layout: German_T5

Table 6.1 Comparison of DNN architectures based on BFP representation

Ref.	Phase	DNN	Exponent selection	Block design	Dataset	Model	Mantissa bits	Accuracy loss %	Area saving %	Power saving %	Speed up (times)
[8]	Training, inference	CNN, GANs	Dynamic, statistics	Filter-based	CIFAR-10, LSUN	AlexNet, WGAN	16	~0	–	–	–
[14]	Inference	CNN	Dynamic, max	Tile-based	ILSVRC	AlexNet	10	~0	–	–	10 ([24])
[3]	Inference	CNN	Dynamic, max	Filter-based	MNIST, CIFAR10	VGG16, ResNet-18, ResNet-50, GoogLeNet	8	<0.3	–	–	–
[12]	Training, inference	CNN, RNN	Dynamic, max	Tile-based	CIFAR-100, SVHN, ImageNet	ResNet, WideResNet, DenseNet	8	<1	–	–	8.5 (FLP16)
[10]	Inference	CNN	Dynamic, max	Tile-based	Sports-1M	Custom	15	0.4	–	92 (Intel i7-950)	8.2 (Intel i7-950)
[7]	Inference	CNN	Dynamic, max	Filter-based	ImageNet	VGG-16, GoogLeNet, ResNet-50	8	<0.14	–	31 (FLP16 [25])	–
[2]	Inference	CNN	Dynamic, max	Filter-based	ImageNet, CIFAR10	LeNet, VGG-16, GoogLeNet, ResNet-50	8	0.12	–	15 (FLP16 [25])	3.76 (FLP16 [25])
[4]	Training, inference	CNN	Dynamic, max	Filter-based	ImageNet, CIFAR100, CIFAR10	VGG16, PreResNet-164, logistic reg.	8	<3	–	–	–

(continued)

602619_1_En_6_Chapter-print ☑ TYPESET ☐ DISK ☐ LE ☑ CP Disp.:26/8/2023 Pages: 94 Layout: German_T5

Table 6.1 (continued)

Ref.	Phase	DNN	Exponent selection	Block design	Dataset	Model	Mantissa bits	Accuracy loss %	Area saving %	Power saving %	Speed up (times)
[21]	Training, inference	CNN	Dynamic, max	Filter-based	MNIST, CIFAR10	VGG16, PreResNet-164	8	0.1	–	–	17 (ARM A53)
[21]	Training, inference	CNN, RNN	Dynamic, max	Filter-based	ImageNet, CIFAR-10, CIFAR-100	ResNet, LENET	≤ 8	<1	92–96	91	–
[5]	Inference	CNN	Static	Filter-based	ImageNet	ResNet 50, VGG 16, Inception, MobileNet	8	<1	~50 (FLP8 MAC)	–	–
[17]	Training, inference	DRL	Dynamic, statistics	Filter-based	–	Q-learning	23	–	15.8	–	–
[6]	Inference	CNN	Static	Filter-based	ImageNet	ResNet-18, ResNet-50	8	<0.6	15 (FXP8 MAC)	16 (FXP8 MAC)	–
[1]	Inference (fine tuning)	CNN, RNN	Static	Tile-based	ImageNet	Many	<=8	<1	88–97	–	–
[18]	Inference	CNN	Static	Filter-based	ImageNet	ResNet 50, VGG 16, Inception, MobileNet	8	<1	~50 (FXP8 MAC)	82 (TITAN GPU)	1.13 (TITAN GPU)
[11]	Training, inference	CNN	Dynamic, max	Tile-based	ImageNet	Custom	3	<2	–	80 (FLP16)	2 (I1)
[13]	Inference	CNN	Dynamic, max	Tile-based	Torchvision, ImageNet	VGG16, ResNet-152	10	<1	50.1 (FLP16 LUTs)	–	1.32 (FLP16)
[20]	Training, inference	CNN	Dynamic, max	Filter-based	CIFAR, ImageNet, WMT14	AlexNet, VGG16, ResNet-18, MobileNetV1, DenseNet-121	Mixed (4,8,16)	<2.31	–	–	1.5 5.3

602619_1_En_6_Chapter-print ☑ TYPESET ☐ DISK ☐ LE ☑ CP Disp.:**26/8/2023** Pages: **94** Layout: **German_T5**

of a specific design choice over the others. However, we can conclude that there is no clear trend in the accuracy enhancement when tile-based blocking is used instead of a filter-based one.

References

1. Darvish Rouhani, B., Lo, D., Zhao, R., Liu, M., Fowers, J., Ovtcharov, K., Vinogradsky, A., Massengill, S., Yang, L., Bittner, R., et al.: Pushing the limits of narrow precision inferencing at cloud scale with microsoft floating point. Adv. Neural. Inf. Process. Syst. **33**, 10271–10281 (2020)
2. Lian, X., Liu, Z., Song, Z., Dai, J., Zhou, W., Ji, X.: High-performance FPGA-based CNN accelerator with block-floating-point arithmetic. IEEE Trans. Very Large Scale Integr. (VLSI) Syst. **27**(8), 1874–1885 (2019)
3. Song, Z., Liu, Z., Wang, D.: Computation error analysis of block floating point arithmetic oriented convolution neural network accelerator design. In: Proceedings of the AAAI Conference on Artificial Intelligence, vol. 32 (2018)
4. Yang, G., Zhang, T., Kirichenko, P., Bai, J., Wilson, A.G., De Sa, C.: SWALP: Stochastic weight averaging in low precision training. In: International Conference on Machine Learning, pp. 7015–7024. PMLR (2019)
5. Fan, H., Wang, G., Ferianc, M., Niu, X., Luk, W.: Static block floating-point quantization for convolutional neural networks on FPGA. In: International Conference on Field-Programmable Technology (ICFPT), pp. 28–35. IEEE (2019)
6. Ni, C., Lu, J., Lin, J., Wang, Z.: LBFP: Logarithmic block floating point arithmetic for deep neural networks. In: IEEE Asia Pacific Conference on Circuits and Systems (APCCAS), pp. 201–204. IEEE (2020)
7. Zhang, H., Liu, Z., Zhang, G., Dai, J., Lian, X., Zhou, W., Ji, X.: A block-floating-point arithmetic based FPGA accelerator for convolutional neural networks. In: IEEE Global Conference on Signal and Information Processing (GlobalSIP), pp. 1–5. IEEE (2019)
8. Köster, U., Webb, T., Wang, X., Nassar, M., Bansal, A.K., Constable, W., Elibol, O., Gray, S., Hall, S., Hornof, L., et al.: Flexpoint: an adaptive numerical format for efficient training of deep neural networks. In: Advances in Neural Information Processing Systems, vol. 30 (2017)
9. Zhao, J., Dai, S., Venkatesan, R., Liu, M.Y., Khailany, B., Dally, B., Anandkumar, A.: Low-precision training in logarithmic number system using multiplicative weight update (2021). arXiv:2106.13914
10. Fan, H., Ng, H.C., Liu, S., Que, Z., Niu, X., Luk, W.: Reconfigurable acceleration of 3D-CNNs for human action recognition with block floating-point representation. In: International Conference on Field Programmable Logic and Applications (FPL), pp. 287–2877. IEEE (2018)
11. Zhang, S.Q., McDanel, B., Kung, H.: FAST: DNN training under variable precision block floating point with stochastic rounding (2021). arXiv:2110.15456
12. Drumond, M., Lin, T., Jaggi, M., Falsafi, B.: Training DNNs with hybrid block floating point. In: Advances in Neural Information Processing Systems, vol. 31 (2018)
13. Wong, Y., Dong, Z., Zhang, W.: Low bitwidth CNN accelerator on FPGA using winograd and block floating point arithmetic. In: 2021 IEEE Computer Society Annual Symposium on VLSI (ISVLSI), pp. 218–223. IEEE (2021)
14. Aydonat, U., O'Connell, S., Capalija, D., Ling, A.C., Chiu, G.R.: An OpenCl deep learning accelerator on Arria 10. In: Proceedings of the 2017 ACM/SIGDA International Symposium on Field-Programmable Gate Arrays, pp. 55–64 (2017)

602619_1_En_6_Chapter-print ☑ TYPESET ☐ DISK ☐ LE ☑ CP Disp.:26/8/2023 Pages: 94 Layout: German_T5

15. Jo, S., Park, H., Lee, G., Choi, K.: Training neural networks with low precision dynamic fixed-point. In: 2018 IEEE 36th International Conference on Computer Design (ICCD), pp. 405–408. IEEE (2018)
16. Das, D., Mellempudi, N., Mudigere, D., Kalamkar, D., Avancha, S., Banerjee, K., Sridharan, S., Vaidyanathan, K., Kaul, B., Georganas, E., et al.: Mixed precision training of convolutional neural networks using integer operations (2018). arXiv:1802.00930
17. Su, J.D., Tsai, P.Y.: Processing element architecture design for deep reinforcement learning with flexible block floating point exploiting signal statistics. In: 2020 Asia-Pacific Signal and Information Processing Association Annual Summit and Conference (APSIPA ASC), pp. 82–87. IEEE (2020)
18. Fan, H., Liu, S., Que, Z., Niu, X., Luk, W.: High-performance acceleration of 2-D and 3-D CNNs on FPGAs using static block floating point. IEEE Trans. Neural Netw. Learn. Syst. (2021)
19. Claici, S., Yurochkin, M., Ghosh, S., Solomon, J.: Model fusion with Kullback-Leibler divergence. In: International Conference on Machine Learning, pp. 2038–2047. PMLR (2020)
20. Noh, S.H., Koo, J., Lee, S., Park, J., Kung, J.: FlexBlock: a flexible DNN training accelerator with multi-mode block floating point support (2022). arXiv:2203.06673
21. Fox, S., Faraone, J., Boland, D., Vissers, K., Leong, P.H.: Training deep neural networks in low-precision with high accuracy using FPGAs. In: International Conference on Field-Programmable Technology (ICFPT), pp. 1–9. IEEE (2019)
22. Fox, S., Rasoulinezhad, S., Faraone, J., Leong, P., et al.: A block Minifloat representation for training deep neural networks. In: International Conference on Learning Representations (2020)
23. Xie, L., He, S., Song, X., Bo, X., Zhang, Z.: Deep learning-based transcriptome data classification for drug-target interaction prediction. BMC Genomics **19**, 93–102 (2018)
24. Zhang, C., Sun, G., Fang, Z., Zhou, P., Pan, P., Cong, J.: Caffeine: Toward uniformed representation and acceleration for deep convolutional neural networks. IEEE Trans. Comput. Aided Des. Integr. Circuits Syst. **38**(11), 2072–2085 (2018)
25. Mei, C., Liu, Z., Niu, Y., Ji, X., Zhou, W., Wang, D.: A 200MHZ 202.4GFLOPS@10.8W VGG16 accelerator in Xilinx VX690T. In: IEEE Global Conference on Signal and Information Processing (GlobalSIP), pp. 784–788. IEEE (2017)

602619_1_En_6_Chapter-print ☑ TYPESET ☐ DISK ☐ LE ☑ CP Disp.:26/8/2023 Pages: 94 Layout: German_T5

Abstract

This chapter examines the use of dynamic fixed point number systems for deep neural networks. We introduce the concept of DFXP and compare it to block floating point systems highlighting their similarities and differences. In addition, we review existing DNN architectures that use DFXP and compare their performance. Additionally, we discuss the various factors that impact DNN performance when using DFXP and explore different approaches for determining the optimal settings of these factors.

7.1 DFXP Overview

DFXP representation shares the same concept of BFP discussed in Chap. 6 and sometimes the notations DFXP and BFP are used interchangeably. As in the case of BFP, in DFXP, the values are grouped and different scaling factors (i.e., shared exponents) are used for different groups. Thus, a scaling factor is unique for each group (e.g., layer). In some cases, it can be changed from time to time (i.e., dynamic). This is compared to the case of FXP which assigns a single global scaling factor for the whole DNN architecture all the time. To this end, Eqs. (6.1, 6.2, 6.3) are applicable to DFXP.

Although several works use the term DFXP to indicate a representation similar to BFP [1–4], the majority of works use DFXP to indicate FXP representation provided with the flexibility to change the place of the decimal point, that specifies the length of the integer and fraction parts for each group of values, Fig. 7.1. This requires that a scaling factor ϵ_N of a group N, (6.2), to be in the range $[-w_N, 0]$, where w_N is the bit-width used to represent elements of a group N [5–7]. Hence, DFXP representation can be reduced to $< I_N, F_N >$ format where I_N, and F_N are the number of bits allocated to the integer and fractional

© The Author(s), under exclusive license to Springer Nature Switzerland AG 2024

G. Alsuhli et al., *Number Systems for Deep Neural Network Architectures*,
Synthesis Lectures on Engineering, Science, and Technology,
https://doi.org/10.1007/978-3-031-38133-1_7

602619_1_En_7_Chapter-print ☑ TYPESET ☐ DISK ☐ LE ☑ CP Disp.:26/8/2023 Pages: 94 Layout: German_T5

Fig. 7.1 The bit representation for the dynamic fixed point number system

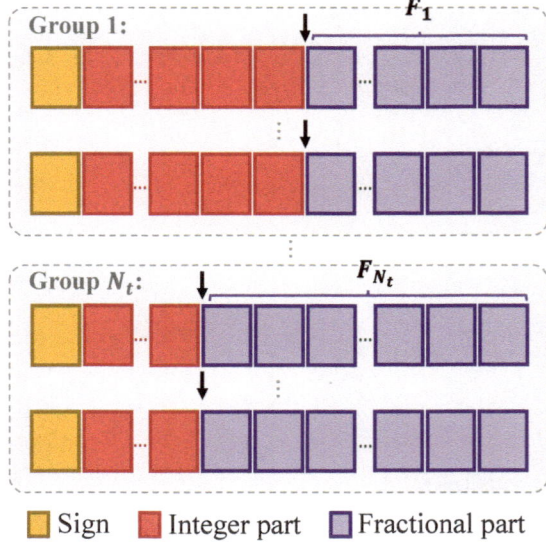

parts, respectively, for all values within a group. Such that $w_N = I_N + F_N$ and $\epsilon_N = -F_N$. Thus, the zero setting problem frequently happens with BFP will not appear for the DFXP at the expense of limited dynamic range, but still better than the one of FXP. We will limit our discussion on these works in this chapter, whereas the other works that use DFXP notation to indicate BFP are discussed in Chap. 6 although they use the term DFXP.

7.2   DFXP for DNNs

A notable difference between DNN architectures that use BFP and DFXP is that the latter gives less attention to the way that the groups (i.e., blocks) are partitioned. The common grouping approach for DNN architecture based on DFXP is to consider the weights, biases, input activation, and gradients vectors (when DFXP is used to accelerate training) for each layer as separate groups and thus associated with different scaling factors [2, 5, 8–15]. Only one architecture presented in [16] statically clusters the filters (i.e., weights) that accumulate to the same output activation of each layer. Then, each cluster represents a group that has its unique scaling factor. The quantization error is effectively reduced with smaller clusters (e.g., when a cluster contains 4 filters) since smaller groups tend to have a smaller range of values.

The main differences between the DFXP representation used in different DNN architectures are the way of finding the best scaling factor F_N and determining the bit-width w_N. The approaches used to optimize the decimal point position and specify the precision of DFXP are classified in the next sections.

602619_1_En_7_Chapter-print ☑ TYPESET ☐ DISK ☐ LE ☑ CP Disp.:26/8/2023 Pages: 94 Layout: German_T5

7.2.1 Group Scaling Factor Selection

The scaling factor (i.e., F_N) assignment to each group in DFXP is usually performed in an offline or online manner.

7.2.1.1 Offline Scaling Factor Selection

The offline assignment is usually used when the architecture is implemented for inference purpose [4–6, 9–14, 16–22]. The common approach for the offline assignment depends on finding the minimum integer bit-width I_N that accommodates the maximum value within a group as in

$$I_N = \log_2(\max(|n_{\max}|, |n_{\min}|)), \tag{7.1}$$

where $n_{\max}$, $n_{\min}$ are the maximum and minimum values within a group N. The remaining bits $w_N - I_N$ are allocated to the fractional part F_N. This approach is used for example in [6, 9]. However, as the presence of outliers in a group results in an unnecessary increase in the integer bit-width, the outliers can be excluded before calculating the bit-width I_N [6]. Several works minimize the impact of the outliers by selecting a scaling factor that minimizes the error between computed and real values [7, 11, 17]. For instance, K-L divergence between FLP32 and DFXP weight distributions is used in [7], whereas a greedy algorithm is utilized in [11] to determine the best scaling factor.

7.2.1.2 Online Scaling Factor Selection

The online scaling factor selection is needed for the training phase in which the values within each group change frequently [1, 7, 8, 10, 14, 23, 24, 24, 25]. Usually, the scaling factor is updated at a given frequency based on the rate of overflow during the training. When the current integer part fails to handle a value in a group, the overflow rate increases. The overflow rate is compared to a threshold to decide whether this scaling factor should be increased or decreased. This threshold can be deterministic and predefined [8, 10, 14, 23, 24], or stochastic [25]. The stochastic thresholding is presented because the lower deterministic threshold results in inaccurate representation for small values while the higher threshold causes large clipping error [25]. The random shuffling between higher and lower thresholds is found to be effective in compensating for the accuracy degradation of the low-precision training (less than 6 bits).

7.2.2 DFXP Precision

The bit-precision of DFXP (i.e., w_N) can be static, mixed, or dynamic with different trade-offs between accuracy and hardware efficiency.

602619_1_En_7_Chapter-print ☑ TYPESET ☐ DISK ☐ LE ☑ CP Disp.:26/8/2023 Pages: 94 Layout: German_T5

7.2.2.1 DFXP with Static Precision

The static precision, which is used in [6, 7, 15, 16, 18, 21, 23], indicates that the number of bits is statically specified prior and is kept fixed for all groups during the training or inference, i.e., $w_{N_i} = w_c$ for $i = 1, \ldots, N_t$, where N_t is the total number of groups associated with a specific DNN architecture. The advantage of this scheme is its simplicity from the hardware efficiency point of view. However, the selected precision is not optimal for all groups, layers, and architectures [25].

7.2.2.2 DFXP with Mixed-Precision

On the other hand, in the mixed-precision scheme, the bit-width, which is determined offline as well, can be different for different groups [5, 16]. The need for mixed-precision mainly comes from the fact that different groups (such as weights and activations) have different required dynamic ranges and thus different required numbers of bits [20]. As the activation results from the convolution accumulation, it is usually allocated more bits. For instance, using DFXP with 4-bit wights and 8-bit activations gives an accuracy degradation within 2% of the full precision using the Resnet-50 CNN model on the ImageNet dataset [16]. In other works, different precision is allocated to different groups in different layers [5, 17]. The authors in [5] stated that a specific fully connected layer activation is more sensitive to bit reduction and it is better to be allocated 16 bits while the activation bit-width of the other layers can be shrunk to 8 bits. This mixed-precision allows them to achieve a 55.64% saving for weights' storage and 69.17% for activations' memory traffic with less than 2.5% loss in accuracy when the Alexnet model and ImageNet dataset are used. The experiments in [17] show similar results. They found that the groups in shallower layers are less robust to bit reduction than the ones in deeper layers. In addition, the computation of the first and the last network layers should use high bit precision to achieve better performance. To optimize the mixed-precision for different groups and to reach the above conclusions, the authors in [17] adopted an iterative bit-precision reduction scheme that aims to discover the groups for which the bit precision can be reduced without causing noticeable performance degradation. When DFXP with mixed-precision is used for training, sometimes different bit-widths are used for the weights during the updates than during the forward and backward propagations [8]. Using higher precision for the weight updates allows for the small changes in the weights to be accumulated precisely.

7.2.2.3 DFXP with Dynamic Precision

The use of DFXP with dynamic precision is presented to adjust the bit-width on-the-fly during training to enable speeding up this process [10, 24, 25]. The scheme in [10, 24] suggests starting with an aggressive initial target bit-width and monitoring the training loss as feedback from the training process. If the training becomes unstable, the bit-width is increased to its maximum value. Afterward, the target bit-width is gradually increased by a unit step for the next trial. This procedure is repeated until reaching the minimum target

602619_1_En_7_Chapter-print ☑ TYPESET ☐ DISK ☐ LE ☑ CP Disp.:26/8/2023 Pages: 94 Layout: German_T5

bit-width that allows for stable training. To maintain the low overhead of this algorithm, it is activated once after each forward/backward computation to find the global bit-width of DNN architecture. A simpler search-based scheme to adapt the bit-width of each layer is suggested in [25]. In this scheme, the convolution is calculated in presence of low and high bit-widths at the same time for several iterations per epoch. If the difference between the high and low precisions is higher than a predefined threshold, the bit-width increases starting from the next iteration till the end of the epoch. After applying this scheme to different datasets and different CNN models, an interesting conclusion was that different datasets require different average bit-widths even if the same model is used.

One added complication of utilizing the dynamic bit-width is the need to design a configurable processing unit that can be configured to compute with various bit-widths during run-time. Thus, the efficiency of the dynamic precision scheme is highly affected by the hardware's supportability of the bit-width levels. Two relatively high bit-width levels (32 bits and 64 bits) are adopted in [10]. The baseline precision to prove the efficiency of their proposed approach is 64 bits which is relatively high training precision compared to other works. On the other hand, [25] could train the CNN with negligible loss of training and testing accuracy using an average bit-width less than 8. This is because they were able to use finer bit-width levels thanks to the bit-slice serial architecture they proposed.

7.3   Summary and Discussion of DFXP-Based DNN Architectures

DFXP and BFP are very similar representations. DFXP can be considered as a subset of BFP with less dynamic range and less hardware complication at the same time. For example, when the DFXP representation is used for CNN inference, the only additional hardware required over the FXP is a simple bit-shifter to align the output activation with the scale factor of the next layer input activation [18, 20]. This simplicity makes it appealing for many DNN architectures [1–28].

By considering the number of accelerators in the literature that utilize each representation, DFXP can be considered the most widely used alternative number system. The popularity of DFXP can be attributed to its simplicity and its implementation in publicly available DNN frameworks like Ristretto [9]. While many DFXP-based DNN architectures have been developed, some have relied on the vanilla DFXP representation without significant modifications. On the other hand, other researchers have explored different strategies for selecting scaling factors and optimizing the bit-width of this representation, which are extensively discussed and compared in the previous sections.

Considering the extensive utilization of DFXP-based architectures (over 30 works), a table summarizing only a few works is included in this chapter for comparison purposes. Table 7.1 provides an overview of these works, evaluating their performance based on metrics such as accuracy degradation, area savings, power consumption reduction, and speed

Table 7.1 Comparison of several DNN architectures based on DFXP representation

Ref.	Phase	DNN	Dataset	Model	#bits	Accuracy loss %	Area saving %	Power saving %	Speed up (times)
[8]	Training, inference	CNN	MNIST, CIFAR-10, SVHN	Custom	10, 12	3	–	–	–
[9]	Inference[a]	CNN	CIFAR-10, ImageNet	LeNet, CaffeNet, BVLC, SqueezeNet	8, FP32	2.3	–	–	–
[10]	Training, inference	CNN	MNIST, Flickr	Alexnet	Dynamic 32, 64	–	31.3 (FXP64 MAC)	45–69 (FXP64 MAC)	5.8 (FXP64)
[11]	Inference	CNN	MINST	LeNet	8, 16, 4	5.9	–	–	–
[23]	Training, inference	RNN	KTH	GRU	24	–	–	78 (FP32 Array)	4.7
[13]	Inference	CNN	–	AlexNet, SqueezeNet, GoogLeNet, VGG-16	8	~0		8 (FXP8 multi-plier)	–
[16]	Inference	CNN	ImageNet	Resnet-101	4, 8	2	–	–	–
[12]	Inference[a]	CNN	ILSVRC	ZynqNet	8	1	–	–	–
[2]	Training	CNN	CIFAR10, CIFAR100	LeNet-5, VGG-16	16	~0	–	–	–
[3]	Training	CNN	ImageNet	ResNet-50, GoogLeNet-v1, VGG-16, AlexNet	16	1	–	–	1.8
[5]	Inference	CNN	Imagenet	Alexnet, VGG16	16, 12, 8, 6	2.5	–	–	–
[26]	Inference[a]	CNN	MNIST	LeNet	4, 8	0.4	–	–	–
[4]	Inference	CNN	–	AlexNet, ResNet-50, MobileNet	8, 16	1.21	–	29.55	–

[a] After training and quantizing using DFXP, these works perform fine-tuning of the DNN weights, by retraining the model, before utilizing them for inference

enhancements. These results are based on the findings reported in the relevant references, allowing readers to gain insights into the comparative performance of different approaches.

Overall, this chapter aims to highlight the widespread adoption of DFXP as a prominent alternative number system in the field of accelerators for DNNs. By showcasing a range of related works and their respective performance evaluations, readers can gain a deeper understanding of the advantages and trade-offs associated with different DFXP-based architectures.

References

1. Sakai, Y.: Quantizaiton for deep neural network training with 8-bit dynamic fixed point. In: 2020 7th International Conference on Soft Computing & Machine Intelligence (ISCMI), pp. 126–130. IEEE (2020)
2. Jo, S., Park, H., Lee, G., Choi, K.: Training neural networks with low precision dynamic fixed-point. In: 2018 IEEE 36th International Conference on Computer Design (ICCD), pp. 405–408. IEEE (2018)
3. Das, D., Mellempudi, N., Mudigere, D., Kalamkar, D., Avancha, S., Banerjee, K., Sridharan, S., Vaidyanathan, K., Kaul, B., Georganas, E., et al.: Mixed precision training of convolutional neural networks using integer operations (2018). arXiv:1802.00930
4. Wu, Y.C., Huang, C.T.: Efficient dynamic fixed-point quantization of CNN inference accelerators for edge devices. In: 2019 International Symposium on VLSI Design, Automation and Test (VLSI-DAT), pp. 1–4. IEEE (2019)
5. de Prado, M., Denna, M., Benini, L., Pazos, N.: QUENN: Quantization engine for low-power neural networks. In: Proceedings of the 15th ACM International Conference on Computing Frontiers, pp. 36–44 (2018)
6. Lin, W.H., Kao, H.Y., Huang, S.H.: A design framework for hardware approximation of deep neural networks. In: International Symposium on Intelligent Signal Processing and Communication Systems (ISPACS), pp. 1–2. IEEE (2019)
7. Guo, J.I., Tsai, C.C., Zeng, J.L., Peng, S.W., Chang, E.C.: Hybrid fixed-point/binary deep neural network design methodology for low-power object detection. IEEE J. Emerg. Sel. Top. Circuits Syst. **10**(3), 388–400 (2020)
8. Courbariaux, M., Bengio, Y., David, J.P.: Training deep neural networks with low precision multiplications (2014). arXiv:1412.7024
9. Gysel, P., Motamedi, M., Ghiasi, S.: Hardware-oriented approximation of convolutional neural networks (2016). arXiv:1604.03168
10. Na, T., Mukhopadhyay, S.: Speeding up convolutional neural network training with dynamic precision scaling and flexible multiplier-accumulator. In: Proceedings of the 2016 International Symposium on Low Power Electronics and Design, pp. 58–63 (2016)
11. Shan, L., Zhang, M., Deng, L., Gong, G.: A dynamic multi-precision fixed-point data quantization strategy for convolutional neural network. In: CCF National Conference on Computer Engineering and Technology, pp. 102–111. Springer, Berlin (2016)
12. Peng, P., Mingyu, Y., Weisheng, X.: Running 8-bit dynamic fixed-point convolutional neural network on low-cost ARM platforms. In: 2017 Chinese Automation Congress (CAC), pp. 4564–4568. IEEE (2017)
13. Lai, L., Suda, N., Chandra, V.: Deep convolutional neural network inference with floating-point weights and fixed-point activations (2017). arXiv:1703.03073
14. Shin, D., Lee, J., Lee, J., Lee, J., Yoo, H.J.: An energy-efficient deep learning processor with heterogeneous multi-core architecture for convolutional neural networks and recurrent neural networks. In: 2017 IEEE Symposium in Low-Power and High-Speed Chips (COOL CHIPS), pp. 1–2. IEEE (2017)
15. Han, D., Lee, J., Lee, J., Yoo, H.J.: A low-power deep neural network online learning processor for real-time object tracking application. IEEE Trans. Circuits Syst. I Regul. Pap. **66**(5), 1794–1804 (2018)
16. Mellempudi, N., Kundu, A., Das, D., Mudigere, D., Kaul, B.: Mixed low-precision deep learning inference using dynamic fixed point (2017). arXiv:1701.08978

602619_1_En_7_Chapter-print ☑TYPESET ☐DISK ☐LE ☑CP Disp.:26/8/2023 Pages: 94 Layout: German_T5

17. Shawahna, A., Sait, S.M., El-Maleh, A., Ahmad, I.: FxP-QNet: a post-training quantizer for the design of mixed low-precision DNNs with dynamic fixed-point representation. IEEE Access **10**, 30202–30231 (2022)
18. Lin, W.H., Kao, H.Y., Huang, S.H.: Hybrid dynamic fixed point quantization methodology for AI accelerators. In: International SoC Design Conference (ISOCC), pp. 282–283. IEEE (2021)
19. Kuramochi, R., Nakahara, H.: An FPGA-based low-latency accelerator for randomly wired neural networks. In: International Conference on Field-Programmable Logic and Applications (FPL), pp. 298–303. IEEE (2020)
20. Prieto, R.N.: Implementation of an 8-bit dynamic fixed-point convolutional neural network for human sign language recognition on a Xilinx FPGA board (2019)
21. Ding, R., Su, G., Bai, G., Xu, W., Su, N., Wu, X.: A FPGA-based accelerator of convolutional neural network for face feature extraction. In: IEEE International Conference on Electron Devices and Solid-State Circuits (EDSSC), pp. 1–3. IEEE (2019)
22. Mitschke, N., Heizmann, M., Noffz, K.H., Wittmann, R.: A fixed-point quantization technique for convolutional neural networks based on weight scaling. In: IEEE International Conference on Image Processing (ICIP), pp. 3836–3840. IEEE (2019)
23. Na, T., Ko, J.H., Kung, J., Mukhopadhyay, S.: On-chip training of recurrent neural networks with limited numerical precision. In: International Joint Conference on Neural Networks (IJCNN), pp. 3716–3723. IEEE (2017)
24. Taras, I., Stuart, D.M.: Quantization error as a metric for dynamic precision scaling in neural net training (2018). arXiv:1801.08621
25. Han, D., Im, D., Park, G., Kim, Y., Song, S., Lee, J., Yoo, H.J.: HNPU: an adaptive DNN training processor utilizing stochastic dynamic fixed-point and active bit-precision searching. IEEE J. Solid-State Circuits **56**(9), 2858–2869 (2021)
26. Lo, C.Y., Lau, F.C., Sham, C.W.: Fixed-point implementation of convolutional neural networks for image classification. In: 2018 International Conference on Advanced Technologies for Communications (ATC), pp. 105–109. IEEE (2018)
27. Yang, J., Hong, S., Kim, J.Y.: FIXAR: a fixed-point deep reinforcement learning platform with quantization-aware training and adaptive parallelism. In: 2021 58th ACM/IEEE Design Automation Conference (DAC), pp. 259–264. IEEE (2021)
28. Kang, Y., Yang, J.S., Chung, J.: Weight partitioning for dynamic fixed-point neuromorphic computing systems. IEEE Trans. Comput. Aided Des. Integr. Circuits Syst. **38**(11), 2167–2171 (2018)

Posit for DNN Architectures

8

Abstract

This chapter discusses the Posit number system, which is proposed as an alternative to the floating point representation. The chapter explains how Posit uses bits more efficiently, allowing for better accuracy with the same number of bits and affording a better dynamic range. The chapter describes how the real number is represented in the Posit format and the common parameters of this representation. It also explains the available approaches for the selection of these parameters and the resultant trade-offs between accuracy and hardware efficiency. The chapter discusses the tapered accuracy of the Posit format, which makes it more suitable to represent normally distributed data efficiently, such as those found in deep neural networks. In addition, the chapter classifies DNN architectures that use the Posit number system and discuss the pros and cons of these architectures. Finally, the chapter highlights the proposed variants of Posit representation to make it more suitable for DNN hardware implementations.

8.1 Posit Number System Overview

Posit number system, also known as type III universal number (Unum) system [1–3], is a floating-like format that is proposed to overcome several shortcomings of FLP representation [2]. Compared to FLP, Posit uses the bits more efficiently (allows for better accuracy with the same number of bits) [4], and has better accuracy and dynamic range [2, 5]. Figure 8.1 illustrates the Posit representation. The w bits Posit number representation consists of four fields; a sign (1 bit), a regime (of variable length $rs \in [1, w - 1]$), an exponent e (unsigned integer

© The Author(s), under exclusive license to Springer Nature Switzerland AG 2024
G. Alsuhli et al., *Number Systems for Deep Neural Network Architectures*,
Synthesis Lectures on Engineering, Science, and Technology,
https://doi.org/10.1007/978-3-031-38133-1_8

602619_1_En_8_Chapter-print ☑ TYPESET ☐ DISK ☐ LE ☑ CP Disp.:26/8/2023 Pages: 94 Layout: German_T5

Fig. 8.1 The bit representation
for Posit number system

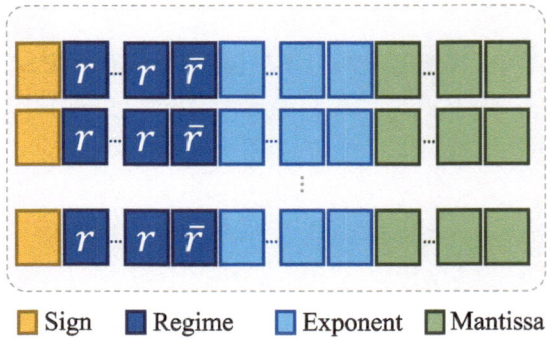

☐ Sign   ■ Regime   ■ Exponent   ■ Mantissa

of fixed length es) and a mantissa (of variable length $ms = w - rs - es - 1$). The regime field contains d consecutive identical bits and an inverted terminating bit (i.e., $rrr \ldots \bar{r}$).[1]

The numerical value of a real number n is represented in the Posit format (by $\hat{n}$) as follows:

$$\hat{n} = (-1)^s \times u^k \times 2^e \times \left(1 + \frac{m}{2^{ms}}\right), \tag{8.1}$$

where s, e, and m are the values of the sign bit, exponent and mantissa, respectively. The used u, and k are calculated in (8.2) and (8.3) with the same order.

$$u = 2^{2^{es}}, \tag{8.2}$$

$$k = \begin{cases} -d, & \text{if } r = 0. \\ d - 1, & \text{if } r = 1, \end{cases} \tag{8.3}$$

Posit representation is commonly characterized by two parameters, mainly w and es, and defined as $\text{Posit}(w, es)$ [1, 7]. The parameter es is used to control the trade-off between the precision and the dynamic range [1]. When the Posit is intended to be used for DNN, these parameters are usually specified in an offline manner regardless of whether this architecture targets DNN training or inference [1, 2, 7, 8]. The selection of these parameters is done usually by experimenting with different parameters and selecting the parameters that give the best accuracy [8] or the parameters that offer the best balance between accuracy and hardware efficiency. For instance, when the exponent length is set to $es = 1$ in [9] a better trade-off between accuracy and energy-delay-product is obtained for $w = 7$ and $w = 5$. On the other hand, the author in [7] decided to eliminate the exponent part (i.e., $es = 0$) as the Posit, in this case, better represents the dynamic range of the used DNN weights.

There are two main differences between Posit and FLP representations, as illustrated in Figs. 8.1 and 3.1. The first difference is the presence of the regime field, and the second is the variability of the mantissa bit-length. Indeed, the innovation in the Posit format comes

---

[1] This is the general case when $rs < (w - 1)$. Otherwise, the regime pattern can be $rrr \ldots r$ when it is terminated by the end of the w bits [6].

602619_1_En_8_Chapter-print ☑TYPESET ☐DISK ☐LE ☑CP Disp.:26/8/2023 Pages: 94 Layout: German_T5

Table 8.1 An example that compares the accuracy and dynamic range of FLP and Posit of the same number of bits. Posit has a wider dynamic range than FLP

	Smallest representable number	Largest representable number
FLP32	1.18×10^{-38}	3.4×10^{38}
Posit(32,4)	3.2×10^{-145}	3.1×10^{144}

from its ability to allocate more bits to the mantissa when the represented number is very small (i.e., higher precision) and fewer bits for large numbers (i.e., larger magnitude) without changing the total bit-width of the format [4]. The Posit is usually known for its *tapered-accuracy*, i.e., small magnitude numbers around the '1' have more accuracy than extremely large or extremely small numbers [10]. An example that compares the smallest and largest representable numbers for FP32 and Posit(32, 4), which require the same number of bits, is shown in Table 8.1. It is noticeable that Posit has a much wider dynamic range and better accuracy near '1' than FP.

The authors in [1] compared the decimal accuracy ($\log_{10} \| \log_{10}(\frac{\hat{x}}{x}) \|$, where x is the actual real number value and $\hat{x}$ is the mapped value according to a specific number system [3]) of different Posit representations to the FLP8 and FXP8. Their experiment showed that: (i) the FXP representation has a peak decimal accuracy so it is suitable to best represent data with a narrow range and uniform distribution, (ii) the floating point has unsymmetrical tapered accuracy (it has gradual underflow from the left and the decimal accuracy falls off a cliff on the right). Otherwise, FLP has almost constant decimal accuracy and it is better to be used to represent data that are uniformly distributed in terms of $\| \log_{10}(\frac{\hat{x}}{x}) \|$ to exploit its efficiency, (iii) and the Posit has symmetrically tapered accuracy which makes it the best to represent the normally distributed data efficiently. Since data in DNNs usually are normally distributed, see for example Fig. 6.5, Posit is expected to be the most attractive number system for DNN [1].

8.2 DNN Architectures Based on Posit Number System

DNN architectures that use the Posit number system usually either rely on the Posit format from end-to-end [4, 5, 11–13] or partially utilize this format and conversion from and to other formats is required within this architecture [7, 14]. These two approaches of using Posit are discussed next. In addition, to increase the efficiency of the Posit number system for DNNs, several Posit variants are proposed. These variants are reported below as well.

8.2.1   End-to-End Posit-Based Architectures

When DNN data are represented in Posit from end-to-end new hardware that is able to perform all operations on these data must be used. In this case, the most fundamental arithmetic operations that need to be carefully designed in hardware are MAC operation and activation functions [4].

8.2.1.1 Posit-Based MAC

Different designs of the Posit-based MAC (or multiplier) are proposed in [1, 2, 9, 13, 15–19]. In most of these works, the MAC design mainly follows the standard FLP MAC as in [1, 2, 9, 16–19]. The main additional steps over the FLP MAC design are the decoding to extract Posit fields of the operands and encoding the result to Posit format [13]. Indeed, Posit MAC hardware implementation is more complicated and less efficient than the FLP MAC with the same number of bits because of the length-variability of the regime and mantissa fields. It is shown in [15] that Posit(32, 6) multiplier has 78% more area and consumes 94% more power than the FLP32 multiplier. This is attributed to the fact that the multiplier should be designed to handle the extreme lengths of mantissa, which is $w - es - 2$, and regime, which is $w - 1$. In addition, the critical path of this Posit multiplier is found to be longer than FLP32 due to the sequential bit decoding required for Posit. By making the fields of Posit format fixed, the area and power efficiency increased by 47, and 38.5%, respectively, over the variable length fields Posit at the expense of negligible accuracy loss. Similar results are shown in [20] as well.

Alternatively, to design a more power and area-efficient multiplier, the authors in [13] proposed a Posit-LNS-Approximate multiplication. This combination allows for exploiting the advantages of Posit accuracy and LNS hardware efficiency. The general concept of performing LNS multiplications is similar to Mitchell's approximation discussed in Sect. 4.3.2, however, by considering Posit format instead. For example, the logarithm of a Posit number is given in (8.4) by taking the logarithm of both sides of (8.1) and applying the approximation in (4.6).

$$\log_2(|\hat{n}|) = 2^{es} \times k + e + \frac{m}{2^{ms}}. \tag{8.4}$$

Consequently, Posit multiplication is performed using fixed point addition. The experiments in [13] showed a significant reduction in the multiplier area by 72.86%, power by 81.79%, and delay by 17.01% compared to Posit multipliers in [12].

8.2.1.2 Posit-Based Activation Functions

The implementation of several activation functions of Posit represented data is discussed in [1, 11, 21]. The Sigmoid activation function in (4.10) is found to be easy to be implemented in hardware for Posit represented data [3]. A few simple bit-cloning and masking are adequate

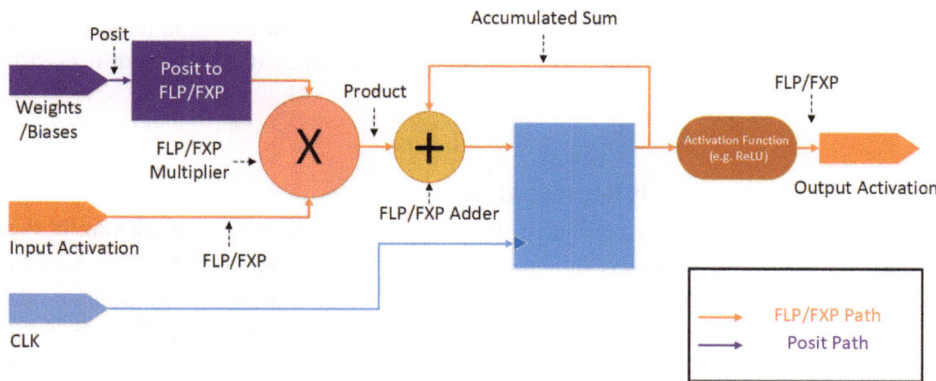

Fig. 8.2 MAC unit for partial posit-based DNN architectures

to approximate this function. Similarly, a fast implementation of the tanh and the Extended Linear Unit (ELU) activation functions are presented in [21, 22], respectively.

8.2.2 Partial Posit-Based Architectures

Several architectures aimed to benefit from the high accuracy and dynamic range of Posit while avoiding its hardware inefficiency by representing only the weights with Posit prior to the inference process [7, 14]. This enables a significant decrease in both the storage and communication overheads. These weights are then converted back to another format, such as FLP in [7] or FXP in [14], during the computation. As we observe in Fig. 8.2, which shows the MAC unit for partial posit-based DNN architectures, the only overhead over the hardware of the standard architectures is the module to convert from Posit to the other formats (FLP or FXP). The penalty of converting Posit to FXP is the increase in critical path delay and power consumption of the MAC by 22.8 and 5%, respectively [14].

8.2.3 Posit Variants

Two Posit variants are proposed for DNNs; the fixed-Posit representation [15], and the generalized Posit representation [6, 10].

8.2.3.1 Fixed-Posit Representation

As its name indicates, the fixed-Posit representation proposes using a fixed length of the regime $rs = constant$ instead of using a variable length in the vanilla Posit. Although the dynamic range and the accuracy of this representation are expected to be less than that of

Posit, using this representation results in much more efficient hardware, in terms of power, area, and delay, with negligible loss in classification accuracy (0.12%) when it used for ResNet-18 on ImageNet [15].

8.2.3.2 Generalized-Posit Representation

The generalized Posit representation [6, 10] proposed a modification to the vanilla Posit format to better represent the dynamic range and data distribution of DNNs. They relied on the fact that Posit with $w < 8$ and a specific es is observed to be unable to accommodate the variability in parameter distributions and dynamic range of different DNN layers and various DNN models. Instead of using mixed-precision Posits which requires a very large search space (as huge as 4^{110} for ResNet-110 when 4 different w values are searched [6]), the Posit format is modified by inserting two hyper-parameters that can be adjusted per-layer to enable a parameterized tapered accuracy and dynamic range. These two hyper-parameters are the exponent bias and the maximum regime bit-width that can be applied by replacing e in (8.1) with $e + sc$, where sc is the exponent bias, and restricting the number of bits allocated to the regime $rs \leq rs_{\max}$. The exponent bias is used to scale the zone of maximum accuracy (i.e., minimum and maximum magnitude values) downward or upward in order to track the data distribution of different layers. The maximum regime bit-width $rs_{\max}$ controls the maximum and minimum Positive representable values. When $rs_{\max} = 1$, the generalized Posit becomes an FLP-like format, whereas it turns into a vanilla Posit format with $rs_{\max} = w - 1$. Various tapered-precision representations can be obtained by selecting the $rs_{\max}$ between these two bounds. The experimental results on several datasets and CNN models showed that the generalized Posit offers considerable accuracy improvement when $w < 8$ bits compared to the vanilla Posit at the expense of a relatively moderate increase in energy consumption.

8.3   Summary and Discussion of Posit-Based DNN Architectures

Posit representation can be considered a variant of FLP. This representation offers better accuracy and a wider dynamic range than FLP. Thus, Posit can represent DNN data more efficiently with the same number of bits. However, in general, the hardware implementation of Posit is found to be more complicated compared to FLP hardware, as it relies on the FLP hardware in addition to the hardware needed to convert from and to FLP. Several trials have been made to enhance Posit hardware efficiency discussed above such as combining Posit with other representations (FXP and LNS) or modifying Posit by fixing or limiting the regime field.

References

1. Lu, J., Fang, C., Xu, M., Lin, J., Wang, Z.: Evaluations on deep neural networks training using posit number system. IEEE Trans. Comput. **70**(2), 174–187 (2020)
2. Carmichael, Z., Langroudi, H.F., Khazanov, C., Lillie, J., Gustafson, J.L., Kudithipudi, D.: Deep positron: a deep neural network using the posit number system. In: 2019 Design, Automation & Test in Europe Conference & Exhibition (DATE), pp. 1421–1426. IEEE (2019)
3. Gustafson, J.L., Yonemoto, I.T.: Beating floating point at its own game: posit arithmetic. Supercomput. Front. Innov. **4**(2), 71–86 (2017)
4. Cococcioni, M., Rossi, F., Ruffaldi, E., Saponara, S.: A fast approximation of the hyperbolic tangent when using posit numbers and its application to deep neural networks. In: International Conference on Applications in Electronics Pervading Industry, Environment and Society, pp. 213–221. Springer, Berlin (2019)
5. Romanov, A.Y., Stempkovsky, A.L., Lariushkin, I.V., Novoselov, G.E., Solovyev, R.A., Starykh, V.A., Romanova, I.I., Telpukhov, D.V., Mkrtchan, I.A.: Analysis of posit and Bfloat arithmetic of real numbers for machine learning. IEEE Access **9**, 82318–82324 (2021)
6. Langroudi, H.F., Karia, V., Gustafson, J.L., Kudithipudi, D.: Adaptive posit: parameter aware numerical format for deep learning inference on the edge. In: Proceedings of the IEEE/CVF Conference on Computer Vision and Pattern Recognition Workshops, pp. 726–727 (2020)
7. Langroudi, S.H.F., Pandit, T., Kudithipudi, D.: Deep learning inference on embedded devices: fixed-point versus posit. In: Workshop on Energy Efficient Machine Learning and Cognitive Computing for Embedded Applications (EMC2), pp. 19–23. IEEE (2018)
8. Montero, R.M., Del Barrio, A.A., Botella, G.: Template-based posit multiplication for training and inferring in neural networks (2019). arXiv:1907.04091
9. Carmichael, Z., Langroudi, H.F., Khazanov, C., Lillie, J., Gustafson, J.L., Kudithipudi, D.: Performance-efficiency trade-off of low-precision numerical formats in deep neural networks. In: Proceedings of the Conference for Next Generation Arithmetic 2019, pp. 1–9 (2019)
10. Langroudi, H.F., Karia, V., Carmichael, Z., Zyarah, A., Pandit, T., Gustafson, J.L., Kudithipudi, D.: ALPS: Adaptive quantization of deep neural networks with generalized posits. In: Proceedings of the IEEE/CVF Conference on Computer Vision and Pattern Recognition, pp. 3100–3109 (2021)
11. Wan, Z., Mibuari, E., Yang, E.Y., Tambe, T.: Study of posit numeric in speech recognition neural inference. Harvard University, Cambridge, MA, USA, Technical Report CS247r (2018)
12. Murillo, R., Del Barrio, A.A., Botella, G.: Customized posit adders and multipliers using the FloPoCo core generator. In: IEEE International Symposium on Circuits and Systems (ISCAS), pp. 1–5. IEEE (2020)
13. Murillo, R., Garcia, A.A.D.B., Botella, G., Kim, M.S., Kim, H., Bagherzadeh, N.: PLAM: a posit logarithm-approximate multiplier. IEEE Trans. Emerg. Top. Comput. (2021)
14. Nambi, S., Ullah, S., Sahoo, S.S., Lohana, A., Merchant, F., Kumar, A.: ExPAN(N)D: exploring posits for efficient artificial neural network design in FPGA-based systems. IEEE Access **9**, 103691–103708 (2021)
15. Gohil, V., Walia, S., Mekie, J., Awasthi, M.: Fixed-posit: a floating-point representation for error-resilient applications. IEEE Trans. Circuits Syst. II Express Briefs **68**(10), 3341–3345 (2021)
16. Murillo, R., Del Barrio, A.A., Botella, G.: Deep PeNSieve: a deep learning framework based on the posit number system. Digit. Signal Process. **102**, 102762 (2020)
17. Zhang, H., He, J., Ko, S.B.: Efficient posit multiply-accumulate unit generator for deep learning applications. In: IEEE International Symposium on Circuits and Systems (ISCAS), pp. 1–5. IEEE (2019)

602619_1_En_8_Chapter-print ☑TYPESET ☐DISK ☐LE ☑CP Disp.:26/8/2023 Pages: 94 Layout: German_T5

18. Langroudi, H.F., Carmichael, Z., Gustafson, J.L., Kudithipudi, D.: PositNN framework: Tapered precision deep learning inference for the edge. In: 2019 IEEE Space Computing Conference (SCC), pp. 53–59. IEEE (2019)
19. Podobas, A., Matsuoka, S.: Hardware implementation of posits and their application in FPGAs. In: IEEE International Parallel and Distributed Processing Symposium Workshops (IPDPSW), pp. 138–145. IEEE (2018)
20. Walia, S., Tej, B.V., Kabra, A., Devnath, J., Mekie, J.: Fast and low-power quantized fixed posit high-accuracy DNN implementation. IEEE Trans. Very Large Scale Integr. (VLSI) Syst. (2021)
21. Cococcioni, M., Rossi, F., Ruffaldi, E., Saponara, S.: Fast deep neural networks for image processing using posits and ARM scalable vector extension. J. Real-Time Image Proc. 17(3), 759–771 (2020)
22. Cococcioni, M., Rossi, F., Ruffaldi, E., Saponara, S.: A novel posit-based fast approximation of ELU activation function for deep neural networks. In: International Conference on Smart Computing (SMARTCOMP), pp. 244–246 (2020)

602619_1_En_8_Chapter-print ☑ TYPESET ☐ DISK ☐ LE ☑ CP Disp.:26/8/2023 Pages: 94 Layout: German_T5

Conclusions and Future Directions

9

In this chapter, we summarize some conclusions and highlight several potential future directions and open research issues.

9.1 Conclusions

Deep neural networks have become a vital component for numerous artificial intelligence applications, delivering an exceptional performance that often surpasses human accuracy. However, their high computational complexity and memory requirements make them challenging to implement, particularly on edge devices. To enhance their performance and enable their deployment on edge devices, research has focused on redesigning DNN algorithms and hardware, including their number representation.

The standard floating point representation provides a wide dynamic range, making it suitable for computationally intensive algorithms that require high precision. However, FLP calculations are complex and power-hungry, making them less attractive for DNN implementation. In contrast, fixed point representations offer greater hardware efficiency but at the expense of accuracy degradation. Several alternative number systems for DNNs offer different trade-offs between energy efficiency and acquired accuracy, including logarithmic, residue, block floating point, dynamic fixed point, and Posit number systems.

LNS simplifies the implementation of costly multiplication operations, offering significant savings in area, power consumption, and cost, with some accuracy degradation resulting from logarithmic approximation. This makes LNS a good choice for DNNs deployed on source-constrained devices for accuracy-resilience applications. RNS exhibits inherent parallelism, utilizing parallel computations along separate residue channels. However, designing an efficient RNS-based accelerator requires minimizing the overhead introduced when

© The Author(s), under exclusive license to Springer Nature Switzerland AG 2024
G. Alsuhli et al., *Number Systems for Deep Neural Network Architectures*,
Synthesis Lectures on Engineering, Science, and Technology,
https://doi.org/10.1007/978-3-031-38133-1_9

602619_1_En_9_Chapter-print ✓ TYPESET ☐ DISK ☐ LE ✓ CP Disp.:26/8/2023 Pages: 88 Layout: German_T5

implementing non-linear DNN operations and optimizing the moduli selection and corresponding arithmetic circuits.

BFP strikes a balance between FLP and FXP, offering different trade-offs through design choices such as block size, shared exponent selection, and bit-width choice. Most of the available DNN architectures utilizing BFP achieved negligible accuracy degradation compared to FLP, even with less than 8 bits, with varying levels of speed, power, and area efficiency. DFXP is a subset of BFP, offering less dynamic range and hardware complication. Posit representation provides better accuracy and a wider dynamic range, allowing for a greater reduction in the number of bits compared to FLP implementations with similar accuracy. However, the hardware required to convert Posit numbers to another number system introduces complexity.

This book provides a comprehensive discussion of the impact of alternative number systems on the performance and hardware design of DNNs. It also highlights the challenges associated with implementing each number system and the proposed solutions to address them. By understanding the trade-offs and advantages of each alternative number system, researchers and practitioners can make informed decisions when selecting number representations for their DNN applications.

9.2   Future Directions

In this section, we discuss several issues and opportunities for future research in DNN number systems, including dynamic number representations, hybrid number systems, and utilization of DNN statistics.

9.2.1 Dynamic Number Systems

The main challenge of using low-precision number systems for training DNNs is the dynamic distribution of weights, activation, and gradients during training. In addition, several works show that optimal parameters of the number system (e.g., bit-widths) can be different for different datasets. This makes a dynamic number system (i.e., a number system that can adjust its parameters either offline or during run-time) highly desirable, especially for training DNNs. However, implementing such a system with online adaptation adds complications to the hardware which should be re-configurable to adapt to the changes in the number system format. Several works that adopt a format with a dynamic bit-width, for example, the author in [1] discussed the worthiness of this approach from a software (accuracy and speed gain) point of view. it seems worth investigating the effectiveness of a dynamic number system from the hardware efficiency perspective.

602619_1_En_9_Chapter-print ☑ TYPESET ☐ DISK ☐ LE ☑ CP Disp.:26/8/2023 Pages: 88 Layout: German_T5

9.2.2 Hybrid Number Systems

Several hybrid number systems have been investigated. Some examples of hybrid representations include DFXP with binary FXP [2], DFXP with ternary FXP [3], DFXP with FLP [4], dual DFXP with DFXP [5], FXP with Posit [6], BFP with LNS [7], Posit with LNS [8], and RNS with LNS [9]. Combining two number systems allows for gaining from the benefits offered by both systems. The hybrid representations are found to be more efficient, from a hardware and accuracy point of view, than using each representation separately. More combinations of these representations can be investigated in the future. For example, applying the same concept of BFP (i.e., each block shares the same exponent) to the Posit number system is expected to relieve the hardware complication compared to the vanilla Posit number system.

9.2.3 Utilization of DNN Characteristics

DNNs has special characteristics that should be considered when searching for more efficient representations dedicated to DNNs. For example, the ability of the neural networks to tolerate the noise is exploited in [10] to design an efficient LNS multiplier by reducing the average rather than the absolute error introduced by the multiplier. This results in enhancing the accuracy of DNN instead of ruining it as would be anticipated when using approximate multipliers. Another example of utilizing the noise tolerance of DNNs is using stochastic rounding (i.e., rounding the number up or down at random) when the real number is mapped to a specific representation. This kind of rounding allowed for training DNNs with lower precision when it is integrated with FXP [11–13], BFP [14], Posit [15], or DFXP [16]. Similarly, the ability to cluster DNN data into groups with narrower dynamic ranges gave birth to BFP and DFP representations. Moreover, realizing that DNN data are normally distributed sheds light on the effectiveness of using the Posit number system, which has tapered accuracy. For future work on DNN number systems, these and other DNN characteristics should be paid attention to achieve more efficient representations.

References

1. Zhang, X., Liu, S., Zhang, R., Liu, C., Huang, D., Zhou, S., Guo, J., Guo, Q., Du, Z., Zhi, T., et al.: Fixed-point back-propagation training. In: Proceedings of the IEEE/CVF Conference on Computer Vision and Pattern Recognition, pp. 2330–2338 (2020)
2. Guo, J.I., Tsai, C.C., Zeng, J.L., Peng, S.W., Chang, E.C.: Hybrid fixed-point/binary deep neural network design methodology for low-power object detection. IEEE J. Emerging Sel. Top. Circuits Syst. **10**(3), 388–400 (2020)
3. Mellempudi, N., Kundu, A., Das, D., Mudigere, D., Kaul, B.: Mixed Low-Precision Deep Learning Inference Using Dynamic Fixed Point (2017). arXiv:1701.08978

4. Lai, L., Suda, N., Chandra, V.: Deep Convolutional Neural Network Inference with Floating-Point Weights and Fixed-Point Activations (2017). arXiv:1703.03073

5. Lin, W.H., Kao, H.Y., Huang, S.H.: Hybrid dynamic fixed point quantization methodology for AI accelerators. In: International SoC Design Conference (ISOCC), pp. 282–283. IEEE (2021)

6. Gohil, V., Walia, S., Mekie, J., Awasthi, M.: Fixed-posit: a floating-point representation for error-resilient applications. IEEE Trans. Circuits Syst. II: Express Briefs **68**(10), 3341–3345 (2021)

7. Ni, C., Lu, J., Lin, J., Wang, Z.: LBFP: logarithmic block floating point arithmetic for deep neural networks. In: IEEE Asia Pacific Conference on Circuits and Systems (APCCAS), pp. 201–204. IEEE (2020)

8. Murillo, R., Garcia, A.A.D.B., Botella, G., Kim, M.S., Kim, H., Bagherzadeh, N.: PLAM: a posit logarithm-approximate multiplier. IEEE Trans. Emerging Top. Comput. (2021)

9. Arnold, M.G., Paliouras, V., Kouretas, I.: Implementing the residue logarithmic number system using interpolation and cotransformation. IEEE Trans. Comput. **69**(12), 1719–1732 (2019)

10. Ansari, M.S., Cockburn, B.F., Han, J.: An improved logarithmic multiplier for energy-efficient neural computing. IEEE Trans. Comput. **70**(4), 614–625 (2020)

11. Lin, D.D., Talathi, S.S.: Overcoming Challenges in Fixed Point Training of Deep Convolutional Networks (2016). arXiv:1607.02241

12. Gupta, S., Agrawal, A., Gopalakrishnan, K., Narayanan, P.: Deep learning with limited numerical precision. In: International Conference on Machine Learning, pp. 1737–1746. PMLR (2015)

13. Chen, X., Hu, X., Zhou, H., Xu, N.: FxpNet: Training a deep convolutional neural network in fixed-point representation. In: International Joint Conference on Neural Networks (IJCNN), pp. 2494–2501. IEEE (2017)

14. Zhang, S.Q., McDanel, B., Kung, H.: FAST: DNN Training Under Variable Precision Block Floating Point with Stochastic Rounding (2021). arXiv:2110.15456

15. Cococcioni, M., Rossi, F., Ruffaldi, E., Saponara, S.: A fast approximation of the hyperbolic tangent when using posit numbers and its application to deep neural networks. In: International Conference on Applications in Electronics Pervading Industry, Environment and Society, pp. 213–221. Springer (2019)

16. Gysel, P., Motamedi, M., Ghiasi, S.: Hardware-Oriented Approximation of Convolutional Neural Networks (2016). arXiv:1604.03168

Glossary

ASIC A specialized integrated circuit chip that is designed for a specific application rather than a general purpose use. Often, ASICs that involve substantial building blocks, such as microprocessors and memory blocks, are commonly referred to as a system-on-chip (SOC).

FPGAs Advanced integrated circuits consist basically of a matrix of configurable logic blocks (CLBs) interconnected through programmable interconnects. These CLBs and interconnects can be programmed to perform desired logic functions. The most remarkable feature of FPGAs is their ability to be reprogrammed in the field to fulfill specific application or functionality requirements after their manufacture.

Kullback-Leibler divergence Also called relative entropy and it is a metric that aims to measure how one probability distribution differs from a reference probability distribution.

Fine-tuning of DNN model Means to adjust the parameters (weights) of the DNN model after training. Usually, this is done by retraining the DNN model After applying the quantization to DNN models to recover the accuracy degradation.

Number system defines the mapping between a real number and the bits that represent it.

DNN quantization In general, quantization means to map a continuous infinite set of values to a small finite set of values to minimize the required number of bits. Rounding and truncation are special cases of quantization. Using any number system (conventional or alternative) to represent DNN data involves a kind of quantization and thus results in a quantization error.

Integer arithmetic The integer arithmetic deals with integers only without fractions. Fixed point numbers can be treated as integers that are implicitly multiplied by a scaling factor. This allows the FXP to use the integer arithmetic unit which is the most efficient arithmetic in terms of speed, cost, and power efficiency.

© The Editor(s) (if applicable) and The Author(s), under exclusive license to Springer Nature Switzerland AG 2024
G. Alsuhli et al., *Number Systems for Deep Neural Network Architectures*,
Synthesis Lectures on Engineering, Science, and Technology,
https://doi.org/10.1007/978-3-031-38133-1

602619_1_En_BookBackmatter-print ☑ TYPESET ☐ DISK ☐ LE ☑ CP Disp.:26/8/2023 Pages: 92 Layout: German_T5

Stochastic rounding In the number system context, stochastic rounding means mapping the real number randomly to one of its two nearest values presented in the finite set of values to be mapped to. In other words, the real number can be rounded up or down at random regardless of which is closer.

602619_1_En_BookBackmatter-print ☑ TYPESET ☐ DISK ☐ LE ☑ CP Disp.:26/8/2023 Pages: 92 Layout: German_T5